Saving the world is not easy, but someone has to do it.

Muhammad Faisal Jhandir MD FRCPC RVT

Preface:

World is slowly but surely getting destroyed. There is no question about it. I don't know how much time the World has got but it does feel like that it is running out of whatever time it has got rather quickly. Question is if and what you, I or anyone else can do about it? Everyone seems to have their own ideas and solutions to the World problems and lot of efforts are being made to restore peace and protect the environment but I still see no slowing down of the destruction of human societies and the environment. Even UN cannot seem to do much about anything anymore.

Saving the World is not an easy job but let's face it someone gotta do it. So why not me! But who am I? You have never even heard of me before. Well you can think of me as a time traveler who has traveled from the Stone Age to the 21st century civilization. I have seen it all there is to see in between. My views and approach to some of the major World problems and controversies are very different than the rest of the World. And I think I am right and the whole World is wrong whether it likes it or not. I understand the odds of me being right and the whole World being wrong is very slim, 1 in 7.3 Billion to be exact. However right and wrong has nothing to do with the odds. No matter how many people say it, a wrong does not become right and a right does not become wrong. One Galileo was right and the whole Rome was wrong. Too bad Galileo did not have the courage to take a stand but I do. Similarly after reading this book I expect that a vast majority of you initially will think that I am crazy and I will take this as a complement. In fact there could not be a better complement for me if actually the whole World thought I was crazy. And that will be a testament to the fact how right I am. You will see what I mean with the passage of time. This is not seventeenth century anymore and I am not in Rome. Now people can reason better and there is justice and freedom of speech where I live. Sooner or later and whether you like it or not you will have to agree with me and respect me for opening your eyes. Sooner you accept the truth better it is for you and the World. But I am not that important nor do I want any importance or attention. I stand alone where I stand and I am satisfied with what I already have. I am not greedy for money or fame. I already make more than what 99% of the people in the World make. What I want to do is to save this planet and its residents. So do not doubt my intentions even though many of you will hate me after reading this book. And I can easily live with that, it won't be the first time. What is important is what I have to say to the World. So dear fellow Homo sapiens pay attention to the book and do what I tell you to do. It would be great if you could understand my point of view but I don't expect everyone to be able to do that. So even if you don't get it, accept it and do what you are told. It's better for you. Trust me, I am a professional.

Mind you, English is my third language and this book in not intended to tickle your literary senses but to provide you an insight on some of the major problems planet and Humans are facing right now from my point of view, how you can actually fix them and why you will not be able to unless you listen to me.

Look around you. Do you see madness in the World? Do you see wars and unrest everywhere with no end in sight? Do you see mass human migrations? Do you see the sky or is it behind a haze of smoke already. Middle East has already been destroyed and the fire is spreading to Europe now. Pollution is chocking up the planet and there is not a thing World can do to stop it. Politician all over the World are busy pleasing people and their campaign donors and are oblivious of what really needs to be done. End is just the matter of time. Only thing left to witness is whether human will kill each other first or will they last long enough for the pollution and global warming to finish the job. I must say that humans of the 21st century are at the same time the smartest and the dumbest humans of all the time. They are smart as the scientific discoveries they have made are truly amazing. The airplanes, spaceships, cars, radios, cell phones, social media, TV and ice cream are all wonderful inventions. But they are also the dumbest in the history of mankind as they have stopped using common sense and logic in favor of political correctness to please their own egos and emotions and make belief equality. Human beings have become emotional creatures and though they claim they believe in science and logic yet they want their emotions to be entertained first and foremost. They believe in freedom of speech as long as it is politically correct and does not hurt anyone emotions even if it is based on logic. My message is not meant for most of the humans nor will many understand it any ways. Many will fiercely oppose it and ridicule it to hide their own lack of understanding and to defend their positions, egos, ways and emotions. People know that I don't want anything from you. And my freedom of speech gives me every right to say what I am going to say as the purpose of this book is not to ridicule anyone for sake of ridiculing it. I am hoping to point out many 600 pounds gorillas in the room no one can see. I am just expressing how I see the World where I stand based on my knowledge, experience, observations and predictions. So reign in your emotions and continue with your ways if you don't understand what I am saying. This message was not meant for you any ways. You keep following the path of your forefathers while I cut my own path as I think I am in a better position to lead than the generations before me. I expect only a handful of people if at all to have the ability to understand and follow my message. They are going to be the ones who will embrace logic and wisdom against their egos and emotions. They will accept that they were wrong and will voluntarily change the way they have been living their lives. To those few my arguments will appear obviously common sense and for the rest this book will serve as nothing more than just a warning they shall ignore.

Introduction:

To me fixing or saving the "World" means saving all the planet's basic elements such as air, sunlight, soil and water and all the life forms living on it. Living creatures are simply water and soil elements put together by a very complex way coded in every life forms DNA or genome. Without sunlight, air, water and soil there could not be any life on earth. Even if anyone of these elements gets compromised life as we know it will cease to exist. As you are probably aware that these elements are under threat in many ways and World despite its efforts is failing to protect them.

Currently both the basic elements and all the life forms living on this planet are under multiple threats. Many species of animals and plants have already gone extinct and every day this list keeps getting bigger. There seems to be no stopping this as the World dwindling forest area is under constant pressure from farmers, loggers and developers. Animal biologist and environmentalist are screaming but no one is listening or even if someone heard them no one seem to have any ability to help them much. Gene pool of almost all species is shrinking except for farmed crops and animals. But for sure overall life's "combined gene pool" or the "genetic database" of all life is shrinking at an alarming rate.

All this destruction is being caused by a single animal species known as Homo sapiens aka human beings. Human population has increased significantly in last few hundred years due to scientific discoveries mainly in medical and agriculture fields. World is considered to be over-populated with Humans now. However how I see it human population can swing rapidly toward the other direction as it happens to almost all animals which when get over populated ultimately suffer a drastic set back in numbers such as lemmings or rabbits etc. This is just the law of nature that whenever any animal population exceeds a certain density it will ultimately lead to mass scale deaths due to either viral or bacterial infections or starvation etc. These catastrophic events end up wiping out a very large portion of the population. Humans are not immune to such events at. A simple influenza outbreak today can easily wipe out most of the humans living on this planet in no time. No one has the ability to do a thing if any such epidemic broke out today.

Human population though over all is increasing it is not the case with all races or demographics. When the civilized World realized that humans were getting over populated they started to cut back the number of children they were having. Every educated and environmentally conscious person contributed by practicing birth control. However this led to a somewhat undesired effect. Uncivilized world didn't practice birth control and continued to reproduce at a high rate. So this has led to an uneven population growth. On top of that female education, cultural atmosphere, use of contraception, monogamy and some other variables have further contributed to decline in population growth in most of the civilized World. For example a recent report on Japan showed that its population continues to show negative growth for last three years including 2014. Similarly most of European countries have a negative population growth. North American population is only positive because of immigrants. So without a doubt the civilized World gene pool is shrinking and over the next few generations the demographics will look very different. One important factor to realize is that human female does not reproduce very fast that is she can bear one child every three years on average and can only continue to reproduce to a certain age. So humans' ability to bounce back from these population shifts might not be as dramatic as say lemmings.

Third World even though reproducing at a much faster rate is facing another kind of threat to its gene pool. The gene pool in third World countries is being permanently contaminated by certain viruses which have the ability to cause chronic low grade infections and establish themselves in the population by vertical transmission that is from mother to the fetus. Three common such examples are Hepatitis B, Hepatitis C and HIV. Infection rates for Hepatitis B or C for example have reached over 90% in certain populations. There are treatments available for Hepatitis C but they are far from being accessible to everyone particularly in third World. So far there is no curative treatment for Hepatitis B or HIV. What's more concerning is that infection rates continue to increase due to ongoing improper sterilization techniques of medical and surgical equipment, re-use of syringes, tattoos, piercings and sexual liberation among many other factors. Third World is losing its battle to curb these infections and once established it will be impossible to get rid of these infections from the gene pool as they will keep getting transferred vertically.

So first world is not having enough children and third world gene pool is getting contaminated at an alarming rate. On top of that there are wars being fought and seems like World is heading towards more racial and religious divide and anarchy. Wars as you know have major influence on adult population mainly on men but now increasingly women and children are being dragged into their destructive paths as well.

Viral infections such as Ebola, Bird flu, swine flu, SARS, Influenza always have the ability to wipe out a large population as they have done before and now with air travel being so common and an absolute necessity these infections would spread much faster and become even more destructive and global.

Environmental issues though not a direct threat yet as much but they will play increasing role in promoting diseases such as cancers, COPD, infections, birth defects and could lead to climate changes, droughts, floods, famines, natural disasters and loss of habitat not only for humans but also all other life form on this planet.

In short all natural elements essential for survival of life that is water, air, soil and sunlight are being compromised throughout the planet Earth. For example river Ravi in Pakistan lost all its fauna and flora to industrial waste contamination decades ago. Colon cancer is considered rare in less than 40 years of age but go look at any cancer hospital in countries like Pakistan and you will see that half of the colon cancer patients are in their early twenties. Water bodies in urban areas throughout Pakistan look essentially black with no sign of much life except may be mosquitoes larvae. If you don't believe me go visit cities like

Gujranwala, Sialkot, Qasur, Multan, Lahore, Karachi in Pakistan etc and you will see what I mean. Pakistan is not the only country where this is happening. Air quality continues to decrease throughout the World and again worse in third world countries. Persistent thick haze is hard to ignore in all major cities of Pakistan, India, Indonesia, Malaysia, Middle East and China etc. This obviously is leading to global warming due to greenhouse effect but it will ultimately decrease the potency of sunlight reaching to earth. In these countries Sun actually "sets" and loses its shine almost two hours before it's supposed to set due to thick haze of smoke.

Soil also is being depleted continuously throughout the World due to erosion of course. But also valuable arable soil is being converted to bricks rendering it useless for creating any new life. Brick making is depleting World soil reserves permanently. Soil reserves which are limited are probably the most valuable of any natural resources as all life and its food comes out of soil. Brick making industry is also contributing to air pollution as it uses a large amount of fuel to bake the bricks. As third world is running out of fossil fuel to burn, these crude industries have resorted to burning tires, plastic, industrial waste and even rubber shoes etc. You can spot them from miles away as a black thick smoke continuously comes out of their chimneys.

Most of the third World still relies on these bricks to make houses to accommodate their ever increasing population. Housing societies are erupting like cancer everywhere in the World right now and increasingly forests and agricultural land is succumbing to housing demands. Soil loss is not only threatening human gene pool but also gene pool for all current and even future life forms to come. As long as there would be soil some sort of life will find a way to bounce back as it has done so for over a billion years but once you turn all the soil into bricks then you have successfully cooked up the planet forever. Mission accomplished! End of story. Good job!

In conclusion not only humans but all life form currently present and in future to come is at risk because of what humans are doing mindlessly. I am not the only one who thinks this way; a lot of people share the same beliefs. A lot of resources have been put in place and a lot of measures are being taken to protect the environment. However what I see is a losing battle so far. Despite your efforts you are heading towards more pollution, more diseases, more suffering, more wars, less children and more chaos. It's a losing battle so far and time is running out quickly.

Question is why no one seems to be able to fix things that is protect the environment, curb wars and bring peace and prosperity to the World. Obviously there are major conflicts of interests for the World to consider. For example if you decrease your dependence on fossil fuel then you risk losing many jobs and the World faces a slump of economy. What do you do? What can you do?

How to fix the World is a complicated question with no easy answer. The world truly needs a Mr. Fixit right now. Humans have to take a step back and re-evaluate all their priorities, beliefs and practices to fully understand what's happening and how can things be handled differently. But problem is humas get emotionally attached to their ways of doing things and will simply refuse to see anything wrong with their way of doing things. Most humans will aggressively guard and defend their ways as no one likes being told that they are wrong. So how do you then see your own flaws? Pretend that you are a time traveler and is just visiting the World for few days. This will give you the ability to isolate yourself emotionally from your own fixed and firmed beliefs of what should be considered right and wrong. Perhaps then you can let go of your ego and look at everything objectively and logically and only then you might share my point of view. Or pretend that you are a biologist studying the last remaining species of genus Homo in the Kingdom Animalia. You are studying an animal species known as "Homo sapiens" who are about to go extinct and your job is to save them. Just like any biologist you have no emotions attached with your subjects. Once you get rid of all your emotions and your pre-formed ideas of right and wrong and focus on the task of saving the species only then you may understand my point of view. Otherwise you will get offended yourself very quickly and refuse to accept the logic in my arguments.

The root causes for many of the World problems lie somewhere else and not where the World thinks in my humble opinion. Humans of 21st century have been barking up the wrong tree. Another example though somewhat crude is that it is just like milking a bull. No matter who is milking it and what technique is being used you are not getting any milk out of a bull no matter how long you try.

Humans need to pay attention to their very own biology better if they want to protect the World and all its life forms including themselves. If humans continue to do things they have been doing then very soon they may end up losing a lot of life forms and even put themselves into endangered category. Current political structure is hesitant to tackle many issues heads on due to conflicts of interests and of course fear of losing votes and donors. So how humans should govern themselves needs to be evaluated before any significant change could be achieved. Whatever humans are doing currently is taking them in the wrong direction and if they want to change this direction they will have to be willing to change anything. Otherwise World will continue on the same path, everyone will keep seeing the destruction coming but will feel completely helpless in preventing it. You will get tired of running races to create environmental awareness as awareness can only do so much unless it gets translated into action. Every civil person right now wants to save the World but really doesn't know what he or she can do more than just running a race or perhaps switching to light savers but this is not going to cut it and you know that. You feel helpless but then you justify yourselves by saying what else can I do anyways, I am just one person and then you go back to watching TV.

Before you read any further I want to ask you two questions.

     1. Do you have any firm beliefs so close to your heart that if questioned you will get uncomfortable?
     2. Does your ego prevent you from accepting defeat in an argument even though you know you are wrong?
If you answer yes to any of the above questions then consider yourself warned. This book will make you uncomfortable and will hurt your ego pretty bad. If your answer is no then great read on but I should tell you it's not the case. You almost certainly have many firm beliefs very close to your heart and

you will get uncomfortable when someone questions them for example religion, sexual orientation, gender equality, equal voting rights for everyone etc. These are all beliefs and I know you have your very firm opinions about them based on information provided to you by people you trust. And now you have your ego and emotions attached and your definition of fairness now relies on these beliefs. Very few people have the ability to make up their own minds about these issues and see pass their own egos and emotions. But yet this is exactly what you need to do if you want to continue to grow and improve. If you get fixated on your old ways then you will never be able to move forward. Sometime moving forward means moving backwards as its not forward or backward that matters what matters is which way is simply better and more logical. It might be against your values and beliefs and might be not as much fun and might be completely opposite to your currently accepted societal norms but if it's based on science and logic you should simply accept it and move on with it instead of trying to be stubborn about your old but wrong beliefs. For example when Galileo said Sun does not revolve around the Earth, it disturbed a lot of people emotions and beliefs but that only hampered science. You should believe in logic of science and humanity only. If you continue to rely on your traditions, political correctness, wisdom of your ancestors, religions, cultures then how can you ever improve? I am not saying these are not important what I am saying is you should critically look at every practice including religion like a wise logical human being should and be happily willing to change your ways if a superior argument and a better path is presented to you.

Continuing with a mistake only makes you guilty of committing a bigger mistake. For example "hand shaking" is customary in many cultures and religions but how I look at this practice is that it has been contributing to spread of infections and leading to human suffering and death all over the World as elaborated by recent Ebola outbreak. So why don't you get rid of it? No more shaking hands and transferring infections. It doesn't serve any biological purpose any ways. But yet people will fuss about it if it was proposed just because they are used to it and their forefathers did it. Have you ever considered that if you are following the ways of your ancestors then you are simply following a generation with inferior knowledge! You are walking backwards. Does it suit wise human beings to do that? Where is the logic in this? But again you are emotionally attached to your "ways" and hence you are stuck. I simply want to boycott it based on my principal that "if a practice leads to human suffering then stop practicing it". I rather hug or do knuckle shake or simply smile, wave or nod. You left your caves a long time ago; similarly you need to keep moving out of your comfort zones if you want to survive better. Don't remain emotionally attached to your "caves".

First World has an added responsibility in my opinion as it is the trend setter for the rest of that World. For example an American celebrity say Angelina Julie gets a tattoo. She makes sure she gets new ink and new needle. Its relatively safe procedure so no harm done and it looks good on her. However say a young fan of Angelina in Pakistan wants to be as cool as Angelina and decides to get a tattoo as well. But unfortunately she is not as lucky to get a sterile needle and ends up acquiring an incurable infection such as Hepatitis C for life and not only her life but life of all the children she was to conceive. So even though the American celebrity did not suffer much harm and she did not mean harm to anyone, she promoted a practice which is harmful overall. That is where your actions are louder than your words and that's why being a role model comes with a certain responsibility. By the way just last year a tattoo parlor in Calgary, Canada was closed due to improper sterilization techniques. So if the first World is not safe then one can only imagine what is happening in the rest of the World.

Opinions expressed in this book are based on my observation and experiences. I have tried my best to remain neutral and pay no regards to anyone emotions, political correctness and religious values. I am presenting the reality how I see it and not how anyone wishes what reality should have been or should have been expressed. I am not sugar coating my opinions to please people or to fit into current acceptable norms of any society. I expect that you are mature adults and can handle someone else's opinion even if it's against your beliefs. There is no need to jump up and down about it. No one will be forcing you to accept my views as the truth.

The Elements, Environment and life forms

All life stems from soil, water, air and sunlight which constitute our environment. Humans and all other life form could be destroyed but as long as these four elements are present life will find a way to bounce back. So these things are the most important things worth preserving and one should make every effort regardless of the cost in order to protect them. If you continue with a practice which is contributing to destruction of these elements than that means you part of the problem and are just being shortsighted. There is no amount of money which makes it worth destroying these four elements.

There are multiple environmental efforts which have been put in place to save the environment particularly in the civilized country. However third World countries where this problem is worse are not doing much at all and have little infrastructure to tackle these complex environmental issues. Living vs. preserving is a tough choice and unfortunately people have no options but to live and governments are apparently busy with more important things then the environment. Lack of awareness among public is probably one of the major cause of democratic governments simply not considering it an issue worth addressing seriously. And public awareness is not likely to happen anytime soon or at least not soon enough. For example in Pakistan most of the water bodies around urban areas are already running black. Air is so think with smoke that you cannot even see the sky and sun is persistently behind a curtain of haze. Soil is being contaminated with industrial waste, automobile industry, pesticides, and domestic waste etc. Third World countries are in serious trouble. And unless they start paying attention to the environmental issues and ensuring proper waste management of domestic and industrial waste they are at risk of quickly becoming inhabitable. Of course people will be living there but it would be a miserable existence. It is obvious that the third World has failed and their leaders cannot deal with the environmental issues either because they simply don't have enough know how or they don't have the resources. However environment is not just a local issue it is a global issue and civilized World needs to take over addressing this problem. Perhaps UN should assume environmental responsibility of the whole World and enforce strict environmental laws regarding industrial and domestic waste management, recycling, CO_2 emission, pesticide use and management of water bodies.

Stop cooking up the Planet:

Another issue threatening soil reserve is urban sprawl which is consuming vast parcels of agricultural and forest land to housing societies. In many countries there are no rules which prevent someone from turning an agricultural land to housing scheme and even if there are rules they are not hard to either ignore or bypass. From large highly planned housing societies to unplanned rural housing schemes are all sprouting up everywhere in the World just like a spreading malignant cancer. There is no emphasis to multifamily housing societies due to both lack of technical knowhow and also most of the people are not civil enough to be able to live so close to each other as it does require certain degree of education, civil sense and mutual respect which unfortunately is in serious shortage in the World right now.

And what are most of these houses being built with? They are made with nothing other than the life most silent and deadly enemy; Clay bricks, made by baking soil using anything which burns as a fuel. No one is paying attention to where these bricks are coming from and how are they being built? You cannot destroy our planet in any worse way than by turning its soil into bricks. You are destroying the soil forever. It's not contamination anymore, its permanent killing of the organic matter. For example Pakistan likely has one the richest soil deposits anywhere in the World as it still has more glaciers than rest of the World glaciers combined. Mighty Himalayan rivers brought this very fertile soil down to plains and these soil deposits left behind by the rivers are exactly where these crude ancient brick making factories known locally as a "buttha" are built. Why? Because this is where the most soil deposits are and they are abundant and free. Yes the quality top soil World is running out of is being cooked back into rocks. In my opinion of all the natural resources such as oil, coal, gold etc soil is the most important and most under looked. All food comes out of soil. If Pakistan wanted it could export its soil to countries like Australia and Middle East where soil is considered a rare commodity but even that would not be a wise thing to do as Pakistan will sooner or later need it herself. These Buttha could be easily seen everywhere in the country. Their tall towers spitting up black thick smoke are hard to miss. The other important and overlooked fact is the fuel source they use to bake the top soil. In Pakistan fossil fuel is very expensive and they have practically run out of forests (less than 4% area left) then what are they burning? Well they are burning everything which burns. Garbage, tires, old furniture, agricultural, industrial waste and even rubber shoes are all fair fuels. Everything which could be burnt and is cheap is being burnt. Nobody is monitoring this industry at all. The only news you hear about buttha is the frequent and serious violations of human rights as they practically have whole families as prisoners and slaves which are forced into making these bricks. Adults and kids forced by hunger and threats work day and night to hand make these bricks which are then baked requiring continuous supply of fuel. Cost for each brick is about 3 cents delivered! Even according to Pakistani standards this is a very cheap raw material. Imagine at 3 cents, there is raw material cost, labor cost, fuel cost and transportation cost and profit! They promise to pay about 0.7 cent per brick labor charges but often end up paying lot less than that as they are basically a mafia. If it was in my hands I would have banned brick making yesterday given what it is doing to soil, air, sun, forests and humans.

But if you ban bricks then what will people use to build houses with? There is currently no viable alternative present especially at this price. Wood is very expensive and not available. Cement is the only viable option as the World happens to have mountains of cement and gravel. Cement is currently being used in major cities of Pakistan such as Karachi and Islamabad where there is heavy machinery available to lift the big heavy cement blocks but in most of the country those cement blocks failed. Why did cement blocks didn't work? Nobody knows for sure, they just didn't is the answer you get. I decided to dig deeper into this issue a bit. My research shows it wasn't the cement people didn't like; cement is relatively cheap and does not require any fuel to cure it. It was the size and dimension of the block. Most of the construction is done there by local masons who only know to make simple homes and they are used to working with bricks. Brick is very versatile and easy to work and transport. It's lighter and could be simply tossed up to the second floor where someone else catches it. No need to carry it up the ladder. Cement blocks were a lot bigger and heavier. They were not easy to work with and were too heavy to be tossed up so had to be manually carried up the ladders. So no one liked working with them.

Once I realized that I took a shot at redesigning the cement block and came up with a cement brick which could be tossed and caught and had the same ease of use as a clay brick. My cement brick replaced three standard bricks and that way construction was faster and saved on labor costs. Just like clay bricks, this was handmade and provided much needed employment to abundant labor. I was paying my labor $3 a day and each person was making about 100 bricks per day. So my labor cost was about 3 cents per brick based on minimum wage compared to 0.7 cents paid by the butthas. My total cost was about 10 cents per brick and given each brick replaced three clay bricks I was very close to beating that price. In many cities clay brick is about 5-7 cents each so I was clearly able to beat that price. Another huge advantage of my brick was in saving labor costs as the construction was significantly faster with it compared to clay bricks. Needless to say my brick was instant hit among the local masons. I built many buildings, walls and bathrooms with those bricks. My cement brick (design copy right protected) has built in air pockets which would give it higher thermal insulation properties and will save on heating/cooling cost of the building as well.

I want to put an end to soil destruction and end to life of the current buttha industry. I want them gone cold forever. These buttha could be converted into manual cement factories as far as I am concerned as they do employee a large number of people. But I would like to see a certain minimum wage and better working conditions for the labor. And I must thank TED talks for bringing that awareness to me as I saw one of the talks where a new city in China needed to be built and they ruled out bricks as they would destroy all the top soil. I did write to UN including Ban Ki Moon himself about 5 times regarding this issue. It has been close to a year now and I am still waiting for a reply.

Solar Cookers can save the planet:

In the war against CO2 emission, deforestation and global warming, a very over looked device is a solar cooker. It's a very simple, practical and World saving invention. Using mirrors or any other reflective surfaces sun light is focused onto a cooking pot and this simply cooks food. It really is magic if you didn't tell people how it works. No fire, no emission and no need for fuel or expensive batteries at all. No need to burn gas, coal, wood, electricity or anything and yet food gets cooked perfect. And on top of that you don't even have to attend to the cooking process. It frees you to go do something else while food is being cooked on its own. It's liberating. It will not only counter global warming, save forests, air, and sunlight but will also liberate women. Women in areas like Pakistan, India, and Afghanistan etc who slave on the stove all day will be liberated. Number one reason for lung cancer and COPD in women in third world countries is smoke from cooking stoves. So imagine the impact of such a simple and cheap device in countries where almost no forests are left. One of the main reasons for illegal logging is need for wood for cooking. Pakistan has run out of stuff to burn. Women when not cooking are looking for wood to burn and many women collect cow dung, mix it with husk and make dung patties all day. Once these dung patties dry up they are used for making fire on which food is cooked. In my opinion it should be compulsory for every household to own two solar cookers at home and one at work.

How the model would work in a traditional household is that a woman can put all the raw materials in the pot and give the pot to the man who while working at his shop could simply keep an eye on it while the food cooks on its own. All men have to do is to re-position the cooker towards the sun every 45 minutes and this takes less than 10 seconds. It is so easy that even a 5 year old could easily do it. Women now will be free to go to work or do whatever they please at home. Obviously if it's the woman going to work then man can prepare the food and she can supervise the cooking at work. Family unit is teamwork and just like in any team everyone does different thing. But either way no need to collect wood or make dung patties all day.

Each solar cooker can save up to 2 ton (2000 kg) of wood from being burnt every year. Imagine if every person in the world was by law required to learn and use these cookers how much greener this planet could get.

What is even more amazing is that solar cooking fulfills all eight of the United Nations Millennium Development Goals (UNMDGs). This is acknowledged by UN Secretary

General's report in 2005.

Following are the 8 UNMDGs:

1. To eradicate extreme Poverty and hunger
2. To promote gender equality
3. To achieve universal primary education
4. To reduce child mortality
5. To improve mental health
6. To combat HIV/AIDS, malaria and other diseases
7. To ensure environmental sustainability
8. To develop a global partnership for development

2015 was the target year set by UN to achieve all the above goals but UN failed to achieve its goals which I find very surprising given they knew for last 10 years how to fulfill them!

Solar cookers are easy to build yourself and you can see many DYI models from the internet. All you need is a card board, aluminum foil and glue. You can also buy them online. However I was not happy with any of the designs available and end up designing my own solar cooker. My design improved the efficiency and hence increased the total amount of food which could be cooked in one day. Also I made sure that they were easy to be produced at mass scale without the need for any power tools as I wanted to employ as many people as possible. We made our cookers out of stainless steel sheets. They were all hand made by local skilled workers who used hammer and shears and sitting under the shade of trees created these planet saving devices.

I also increased the size of the these cookers so they can cook food for 10-15 people at a time twice a day compared to other commercially available models which cook food for only 4-6 people at a time. I also made a commercial model which could cook food for 70 people twice a day.

Cooking pot has to be black, its better if it's airtight such as a pressure cooker. Yyou need a heat resistant plastic bag which goes around the pot to trap the heat just like a green house. I had been trying to promote these cookers for many years in Pakistan. Now I have decided to go commercial as there is a huge market and a good profit margin. At the same time I will be giving employment to many people involved in construction, distribution and education. I plan to go with direct sales models where local village women will train others and also act as sales rep. This is a win-win product.

I have also invented a new concept of "Solar Kitchens" on the go. I have figured out a way to turn an ordinary packing box into a solar cooker and then back into an ordinary packing box. This way all the necessary items required to cook and survive on the go could be packed inside the box. When you need to cook food simply take the items out, turn the box into a solar cooker, cook your food, feed your family and then when done turn the cooker back to box and re-pack everything back into the box and off you go . Everything needed to survive including black cooking pots, plates, cups, cutlery, cooking oil, salt and pepper, clear plastic water bottle to be re-used to purify water, hand soap, dish washing soap and sponge, a blanket, rope, first aid kit, multivitamins, flash lights, lighter, a backup grill to cook

Currently there are more than 50 million displaced people on the planet. This 1 cubic ft of ordinary cardboard box could be the difference between life and a miserable death for many of those people. I have decided not get a patent on my design and just like Alexander Fleming gave penicillin as a gift to humanity, I am giving Solar Kitchens as my gift to not only humanity but all life form on this planet.

Here is a quick and practical way to save the World: Every household in the World especially wood burning ones must own at least two of these solar cookers. Make this a law and you have saved the planet overnight and in fact you can make it greener than it has ever been. This will address the dreaded 2C rise in global temperature, deforestation, CO_2 emission, women liberation, poverty etc and provide much needed boost to economies as this industry will create many jobs for manufacturing, distribution, sales and education. And UN can achieve all its UNMD goals as well! You are welcome. Just like seat belts and helmets are compulsory because they save lives, solar cooking must be compulsory too because it saves the Planet.

In my opinion burning fossil fuel could be "beneficial" for the planet if done in a controlled way. A lot of carbon got locked away from the cycle of life in the form of fossil fuel. By burning it and turning it back into carbon dioxide (CO_2)

you are enabling carbon to enter the cycle of life again. All you need to do is to plant forests and stop cutting and burning trees so they can turn this CO2 into organic matter. Learn to differentiate "fossil" fuel from "live" fuel. As long as you only burn fossil fuel and do not burn wood, CO2 emission is a manageable problem. One can set achievable targets by balancing CO2 release with CO2 harvest from the atmosphere. But time is running out for these solar cookers as sun potency is going down fast due to haze and soon this window to save the World will disappear forever.

Solar Kitchen in a box
scale 1 cm = 10 cm

Saving Homo sapiens and other living beings:

There are number of reasons humans are suffering and dying. Obviously I cannot talk about every problem so I have decided to pick the most common problems affecting humans and the ones which are relatively easy to fix. I will start with some common medical problems and discuss the complex and sensitive issues at the end.

Arthritis:

The number one reason humans see a doctor is arthritis, specifically osteoarthritis. What is osteoarthritis? If you are asking this question that means you are not 40 years old yet. Otherwise you would already know what a pain in the neck, back, hips or knees this osteoarthritis is. It's a chronic crippling and painful condition which can affect any joints of human body. Spine hips and knees arthritis is most common and most damaging as it affects your ability to walk and threatens your independence the most. These are all weight bearing joints. Why does arthritis occur? It's simply wear and tear of your joints. With age this wear and tear leads to damage to the cartilage of the articulating surfaces and starts causing pain, then limits range of motion and then ultimately leads to complete failure of the joint. There is no cure for it. Pain is managed by pain killers such as **NSAIDS** like ibuprofen or naproxen which can themselves cause gastric ulcers, hypertension and kidney failure among few other side effects. Once the damage gets to a certain point then you need a new joint. Most of the hip, knee, and spine surgeries are done because of osteoarthritis. In many hospitals in Canada, there is so much demand for these surgeries that one has to wait for over a year to get it done. People travel to other countries and pay out of their pockets as they can't bear the pain and loss of independence. Almost all humans will get this painful crippling condition. You already have it or you will sooner or later. Why are humans prone to osteoarthritis? There are lots of theories such as humans are living longer than when they lived in the caves etc. Your work and life style are also blamed. But none of these explain the burden of this universal disease in my opinion. Why is that that your mind, heart, lungs, stomach, Liver, kidneys and almost every other organ has a life span of about 80 years but not your bones? One can live a textbook life according to doctors' advice and still develop this crippling condition by age 50. In fact by age 30, 50% of humans have some degree of spinal degeneration.

The answer lies in understanding how human body works. A human is nothing but a biodegradable machine with artificial intelligence. As it's a mechanical machine all the laws of physics apply to it. When you stand up straight and ambulate you always maintain your center of gravity right over your feet. Your head is heavy so cervical spine curves a bit forward to adjust for that. Your chest and belly are full of organs which are heavy so lumbar spine also curves forward as well. These forward bends at cervical and lumbar spines are known as lordosis of the spine. The entire spine is supported up right by the muscles behind the spine and the biggest muscle of human body that is gluteus maximus or butt muscles. When a human stands straight up, it is well balanced and center of gravity lies somewhere right over the top of the arch of the foot. The center of gravity has to be maintained closely when ambulating otherwise you will fall forward or backwards. And maintaining this is a work of art. This is the main reason biped robots are hard to make as they are prone to falling. Robots walk like robots simply because it's hard to maintain the center of gravity in other words balance. Humans can do this without thinking at all; you can walk forward backwards, sideways, on one leg and run while always positioning your center of gravity right above your feet or the foot which is in contact with the ground. The most amazing is how you maintain your center of gravity while going up or down. For example you will fall flat on your face going downwards if you didn't lean back to keep the center of gravity right above your feet. Same way you will fall backwards if you didn't lean forward while climbing up. Experiment yourself and see what happens. Go up on a ramp and see how your body will lean forward. When you stand straight on a level ground with arms to the side and palm facing forward without your shoes then your body is in neutral position and every bone is in correct anatomical position in relation to the bone above and below. All the joints angles for example angle between the heel and shin bone, angle between tibia and femur and angle between all the vertebrae etc are exactly where they are supposed to be. Let's call these angles "standing angles". Force of gravity is travelling downwards from one bone to the next through these joints from exactly the points it is supposed to travel.

But what happens to your body when you stand with your shoes on? Should your body position be any different if you are still standing on the same level ground but only now you have put your shoes on? Should the shoes be changing your "standing angles" at all? The answer is obviously no. But do the shoes change your standing angles? The answer is yes. Almost all shoes on the market right now change the standing angles of your weight bearing joints and invariably make the gravitational force travel through different points on the bones then where it was supposed to travel. The reason behind that is that more than 99% of the shoes on the market make your heels sit few degrees higher than your toes creating a heel-to-toe gradient as if you are standing on a downhill slope even though you are not. In order to maintain your center of gravity over your feet your whole body leans back exactly the same way as if you are standing on a downhill slope even though you are standing on a level ground. In high heel shoes it's very obvious but even in running, hiking, military, work, kids and almost all shoes on the market right now there is a randomly selected heel-to-toe gradient. There are very few shoes which maintain the body natural standing angles. Most shoe companies just randomly add a certain amount of heel-to-toe gradient with no consideration at all to how it affects the body standing angles and how it changes the way force travels through your skeleton.

Now if you are say going downhill on a grade of 7 degrees then your shoes may artificially increase that gradient for your body to say 10 degrees or to 40 degrees which depends on how much heel-to-toe gradient is built in your shoes. Foot without the shoe has a zero degree heel-to-toe gradient as the heels sit exactly at the same height as toes. As your height is about 10 times more than the distance between your heel and ball of your foot, 1 cm lift at the heel will result in 10 cm forward displacement of your head and as a result your center of gravity will no longer be over your feet and you will fall forward if not supported. So you have no choice but to lean back in order to bring the center of gravity back on top of your feet. You don't realize this much because your body does this automatically for you. Every 1 cm heel lift increase the heel-to-toe gradient by about 4 degrees for a European size 40 shoe and every 1 inch heel lift increase this gradient by roughly 10 degrees for the same size foot!

A 3 inch heel would add a nearly whopping 30 degree heel-toe-gradient for a European size 40 shoe! This angle would be even steeper for a smaller foot. Wearing that shoe would move a person head exactly 30 inches (two and half feet) forward from its original position if the person was to maintain all its natural standing angles. This will make the person fall flat on its face if the body did not lean back to bring the center of gravity over the feet again. In order to prevent from falling forward, legs and spine move backwards from their original position and angle between the shin and foot joint (tibiotalar joint) increases dramatically and the spine goes into a hyperextension or hyperlordosis. As the standing angles of the weight bearing joints change, gravitation force travels through completely different points now.

When the spine is in hyperlordosis a small vector of force is going anteriorly and not down the spine as it was supposed to. As spine is made up of many individual vertebrae stacked on top of each other this hyperlordosis could make them prone to slipping on each other. This risk gets even worse when one is carrying some load or going uphill or downhill. Then this small forward vector can become big enough to cause damage for example disc herniation, nerve root impingement or chronic degeneration and remodeling of the bone creating bone spurs etc. Overweight people are naturally more prone to it as the spine is already stressed. Same way, hip, knees, ankles and feet all notice the change. Simply put when you change the standing angle of any joint the force will travel through a different point then where it was supposed to and this could cause damage to the bones and joint over time leading to arthritis. This small angel change at heel is not a small change by any means when you are talking in terms of physics. Imagine a statue standing up straight and then imagine someone sliding few coins to raise the heels of the statue. How many coins will it take before the statue falls on its face?

Now question is why elevate heels at all? Was there a purpose for lifting up the heels? Well turns out heels were invented for horse riding around early 11th century in England. Added heels allowed horse riders to control the stirrups better. Because horse riders were considered elite and these shoes gave them a "proud" stance, it became fashionable in men to wear heels. So contrary to the popular belief, boots weren't made for walking; they were made for horse riding.

Later women adopted heels and turned out that this high heels induced hyperlordosis created a posture in women known in biology as "lordosis behavior" or "mammalian lordosis" or "presenting" observed in some animals like female cats and rats when they are in heat or sexually receptive. Hence high heels added to women sexuality by creating this "presenting" posture. But for last 1000 years no one has paid any attention to what long term changes this artificial and persistent "presenting" posture has been doing to bipeds Homo sapiens who naturally do not have a "presenting" posture.

Human body is designed to carry weight almost equal to its own body weight safely. It will simply bend forward at the hip joint while still keeping the spine straight like a flat bed. But when a human wearing shoes with a heel-to-toe gradient is lifting weight then spine will be a little too upright than where it should have been. So instead of gluteus maximus doing most of the lifting as it should, smaller muscles of the spine feel the weight and that could lead to spine damage. So my advice to shoe companies is to start looking for good lawyers as they will need them soon. I can already see all the millions of dollars Worker Compensation Board (WCB) claims going to shoe companies very soon. One could argue given the damage and pain heel lift probably has caused to human body along with cost of healthcare, lost of work and independence, falls, fractures and deaths this is one the most negligent and in fact criminal act of history. Shoes should only cover and support the foot without changing any angles or shape of the toes.

What is the story with pointy shoes? Do feet look like arrowheads to shoe designers? Foot is longer at the big toe and moving outwards almost form a semi-circle and that is how shoes should be. It would be nice if shoe companies hired some engineers and perhaps doctors who can actually study the design of the foot. Someone with working eyes will do too. Simply make the shoes cover the foot to protect it from the elements and thorns and do not distort its anatomy or design.

I do want to warn about something here. As most of you have been wearing shoes with heel-to-toe gradient most of your life then your bone and muscles have gotten used to this abuse. As you stop wearing elevated heels, your bones and muscles which have not been used in the normal way ever will feel all of a sudden an increase in work load. You will feel pain in your feet and calves for some time. Your muscles could feel sore up to few weeks. If you are an athlete or a runner then make sure you respect this fact. Do not and I repeat do not try to do the same intensity of exercise you are used to. Drop down to 10 % or even less and give yourself 6 months at least to get back to the same intensity of workout. Pretend as if you are learning to walk again. Otherwise you will injure your muscles, rupture tendons or may end up with stress fractures of your small bones of your feet. Your shoes size will also likely increase and particularly the foot will get wider and look stronger. Take picture of your feet before and after 6 months and you will see what I mean. I would recommend getting rid of any heel lift all together but then very slowly rebuild your muscles. Some will be able to make this transition faster some will make it slower. Make pain your guide. There will be some pain for sure as there always is when one starts a new workout routine. But use your common sense, be nice to your body, pain should be what you feel when building new muscles and not the injury kind. If in doubt know that you have over done it and decrease your walking or running distance. With time your body will learn to walk and run again the way it was supposed to.

Clothing which restrict the natural movement of human body also alters how human body mechanics work if they limit the natural range of motion of a joint. For example very tight pants will decrease the range of motion of hip and knee joint which will alter the mechanics of walking. This restriction will change how the forces transfer through the body and that overtime can add to wear and tear of the joints. If you already have arthritis or even if you don't you will notice how certain clothes make you tired easily or cause pain in knee or pelvis and how good you feel when you get out of those clothes. You should not wear clothes or shoes which when you get out of them you feel instantaneously better. Your body is trying to tell you something. Wear clothes which are either lose enough or stretchy enough to allow natural range of motion of all your joints. My favorites are stretch jeans and pants. I as a rule will only wear clothes and shoes which I can comfortably run in anytime. Such clothes are comfortable and in life you never know when you will need to run. Life is full of surprises and one should always be prepared to run if needed to for example to save your own or someone else's life. I don't want to be in a position where I have to say that I have to go home to change into my running clothes first. Always be ready to run and have the fitness to run at least 1 km. It's not too much to ask from the most elegant running machine ever made. It might save your or someone else's life someday.

Alcohol

Lot of people love alcohol and can't think of fun without thinking of alcohol. Why is that? Well, one it does temporarily makes life a bit more fun and two you have been programmed to believe in this relationship. Every major sporting event is sponsored by alcohol industry. Clever media advertisements make you believe that without alcohol you cannot have fun. Kids can't wait to get to legal drinking age and when they do most of them let the world know they have arrived. How much one can drink gives one certain bragging right. Most of the college life for many of the students is nothing but drinking alcohol. All parties, sporting events, graduation ceremonies, spring breaks, ski clubs or any recreational sporting clubs are all about alcohol. One could easily predict a future of a student based on how much alcohol one consumes. It will be an inversely proportional graph. More you drink the less likely you will be successful. Unfortunately many otherwise smart people end up wasting their lives this way. Hospitals are full of alcohol damage. Roughly nearly 30% of the patients in any hospital in North America are there because of alcohol. If alcohol was a medical drug which many claim that it is, it would have been banned a long time ago as the damage it causes to humans is way more than any potential benefit it provides if any at all. If you go to any hospital, go visit the trauma center, you will find alcohol role in most cases, go to medical wards and you will see how beyond repair are the bodies of alcoholic cirrhotic are, go to neurology and you will find the same story. No matter where you go you will see alcohol victims. So many car accidents, so many broken homes, so many broken children abused by their own parents, so much human misery and suffering could be attributed to this one drug. Yet it doesn't even have a warning on its label. You blame the patients for letting them get damaged by it but you don't take the public health approach with it. I have seen very smart people including engineers, doctors, executives and scientists getting destroyed by this drug. No one is safe from this silent killer. Most of the alcohol damaged patients are regular people who are not "alcoholics" by definition. They didn't think they were over-doing it. But they got damaged, they got blind sighted. Society as a whole has turned a blind eye to alcohol damage. Somehow no one can see this white elephant in the room. May be there is too much money at stake or may be people like it too much. How many bodies in a river one would need to see before putting up a no swimming sign or at least a warning of any kind?

What kind of toxin alcohol is? It is a cytotoxin, neurotoxin, hepatotoxin, hemotoxin and myotoxin. It kills neurons (brain cells) on contact. Here is something you might find interesting. When doctors need to destroy a nerve completely and permanently for reason such as nerve impingement or to block the pain conduction etc., what do they inject in that nerve? Yes you guessed it, its alcohol and the procedure is called alcohol ablation. When injected inside a nerve it kills that nerve permanently. Alcohol destroys liver cells and is the leading cause for permanent liver damage called cirrhosis in the civilized World. Its hemotoxin and you can see its effect on blood smear right away as it suppresses maturation of all cell lines. You can see its damage in almost all organs of the body from head to toe. It can cause loss of vision, irreversible memory loss, loss of balance, cardiomyopathy, esophageal cancer, gastritis, pancreatitis and pancreatic cancer, liver cirrhosis, bone marrow suppression and death. Name almost any disease and alcohol is up there as a likely culprit. It is a solvent and just because you dilute it, mix it with fruit juices and give it fancy names does not change its chemistry. It's damage is additive. If you drink daily or on most days then sooner or later you will notice its effects. I have seen damage from two drinks a day. How did doctors even come up with that recommendation? It's misleading and data supporting this recommendation to be safe is simply not there. Its vague, doesn't take into account how long someone has been taking it, doesn't take into account the body size or ethnicity etc. It is a poison and diluting it makes it only a diluted poison. Drinking it less makes it a case of slow poisoning only. It remains a poison and has poisoned many lives. Why allow poisoning at all? Why promote it? Why let your youth get destroyed by it? You should at least treat this the same way as you treat tobacco. Would you put a recommendation on cigarettes for example saying 5 cigarettes a day is okay? No it's not okay. Smoke is damaging to lungs. Smoke cooks meat so basically smokers are cooking their bodies from inside out. Same treatment should be given to alcohol. Only safe recommendation is to stay away from it. Less poison in the body is still poison in the body. And the much inflated cardiac benefit is not proven and even if it was proven then it gives alcohol a therapeutic drug status and then you should subject it to same safety regulations as all therapeutic drugs are subjected to. And I will see what happens to it then. It will get destroyed; it will never ever pass the safety requirements set by FDA, given its damaging effects on every organ on the body. This therapeutic benefit is thrown in just to confuse the issue. Let's decide this. Is it a drug or not for once? So lets ask FDA to weigh in on alcohol. Is it a drug and should doctors be even giving any kind of recommendations at all? Has the 2 drinks per day dose been rigorously tested with randomized double blind controlled long term studies? The answer is no. So there is no long term safety data for this recommendation and yet there is plenty of harm data. Go to any hospital or any other place where suffering humans go and you will see why I am so against this drug. How much human misery a single drug can cause is just unreal.

Infections:

Infections have been a major enemy and have caused much suffering and death throughout human history. Before the discovery of antibiotics humans were quite helpless against bacterial infections. With the antibiotics humans have mostly won the war against bacterial infections. Infections which had very high mortality rates such as meningitis, endocarditis, and pneumonias are now easily curable. However humans are losing some of their ammunition as bacteria continue to evolve and are becoming increasingly more and more resistant to antibiotics. Many resistant bacteria has been reported such as MRSA (Methicillin Resistant Staph Aureus), VRE (Vancomycin resistant Enterococcus) and more recently NDM1 (New Delhi Mutation 1). For example NMD1 gives otherwise easily treatable bacteria such as E.Coli a great advantage as it has developed resistance to one of the most potent and go to drug known as carbapenems. It's been reported almost all over the World now. Another increasingly common resistant organism is ESBL positive E.Coli. A simple E.coli urinary tract infection which is usually treated easily with three days of oral ciprofloxacin at home, total cost for simplicity say is at most $100 dollars including the doctor fees. But if this E.coli happens to be ESBL positive then you are looking at two week of hospitalization for IV antibiotics.

Cost for single day in the hospital in Canada for in-patient treatment is roughly over $3000 dollars so this UTI will cost close to $50,000 of taxpayers' money to treat. So you can see how that changes the game. What's more worrisome is that antibiotics resistance is on the rise and it seems like a losing war. It's just a matter of time and you will soon have bacteria which will be resistant to all the current antibiotics. For example there is already MDR (Multi drug resistant) Tuberculosis which is resistant to all known antibiotics. So what does that mean? It means when your kids will get pneumonia you might not have any antibiotics left to treat them with. You are going back to pre-antibiotic era. But there might be hope. You are as always optimistic that you can come up with new antibiotics. Well even if you do these antibiotics will likely be very expensive as not everyone is as generous as Alexander Fleming was. He never filed for a patent on penicillin and simply said that this was his gift to humanity. It was a selfless move but maybe he should have as humans don't respect free things. And perhaps they would not be abusing these antibiotics and taking them for granted if he in fact had filed a patent. But let me assure you that new antibiotics if and whenever get discovered will have patents and will have a bigger price tag as the current pharmaceutical industry works for profit mainly. How potent and effective these antibiotics will be and what kind of side effects they will have is again nobody knows.

But what's more important is to preserve these antibiotics for your coming generations and limit their widespread abuse. These antibiotics should not be the domain of any pharmaceutical companies as they belong to all humans. Pharmaceutical companies promote their antibiotics use very heavily in the third World as there are no rules on industry involvement and relationship with doctors. Many doctors are paid kickbacks in forms of overseas trips, office overheads, supplies or even cash for ordering these antibiotics. Also lack of easy microbiology labs in third world where a doctor could identify the bacteria and its sensitivities, leads to simply prescribing the most broad spectrum and "potent" antibiotics. And marketing strategy of simply "this will kill any bacteria" makes sense to patients and doctors alike. First world needs to step up and help out the third world by opening "Infection diseases control centers" to fight this war. These centers will be relatively cheap to build as all they need is a good lab and few outpatient clinics. People like Bill Gates need recognition for their work in this fight. Third world is losing its battle to infections. NDM1 was discovered in India but now has become a global problem. So even though you might think that you are safe as you are in first world, this world is a very small place. And with air travel being so frequent there are really no boundaries at all for infections.

In contrast to bacteria humans are still mostly helpless against viral infections and are vulnerable to epidemics capable of wiping out a large population in a short amount of time. Good examples are influenza epidemics such as Spanish flue which killed millions less than a century ago. Recent epidemics such as Ebola and H1N1 expose humans' ongoing vulnerability to such epidemics. If anything these epidemics will be more devastating given frequent air travel. If you think current medical advances safe guard you from these infections than think again. These viral infections are like a horse with no reigns. You may or may not be able to do something against them despite all your technological advances. Third World countries are especially vulnerable as they don't even have the same "isolation" abilities. For example in Pakistan hospitals didn't even have a basic face mask with N95 rating which can block viruses such as H1N1. There are reports of doctors not seeing patient suspected of H1N1 and I don't blame them. Doctors are humans and are equally vulnerable to such infections. And they will be the first to die off if not protected by the proper equipment.

So what can you do better than you are already doing? How can you decrease the chances of spreading these air-borne infections? I have few suggestions. They are simple common sense things but they could have a major impact. First and most important is hand washing as you all know it. But question remains how often? It depends just like for any drug on the "half life" of clean hands. Believe it or not "half life" of clean hands is very short. Many times it's as short as few seconds. For example when you touch the tap to shut of the water after washing hands, you just re-contaminated your hands, you grab the handle of the bathroom door and there your hands are no longer clean. And technically you need to wash your hands again. This degree of compulsion though may not be practical but it doesn't mean it's not needed. So you somehow need to increase the half life of your clean hands. How can you do that? Here are few simple tips.

1. After washing hands don't touch the taps, automatic taps are great for that reason.

2. Bathroom doors should open outwards or could be pedal operated or you can use the paper towel to open it. I will use my clothing to grab the handle if there are no paper towels

3. Do not shake hands - this custom should be abandoned as it leads to spread of infections as recently elaborated by Ebola outbreak. As a rule human should simply abandon any practice which causes human suffering.

4. Do not share your pens with anyone. Always carry your own pen

5. Do not touch the light switches or elevator buttons as they are very dirty. I use my elbows to turn them on and off

6. Keep your hands in your pocket when you are not using them

7. Avoid touching other common or shared things such as magazines in doctors waiting area as they are dirty, peanuts in a bar, any door handles or any other object touched by someone else etc. And if you do then be aware that your hands have become dirty and will need cleaning.

8. TV remotes and telephones etc in the hotel rooms and airplanes are dirty beyond belief

9. Keep a hand sanitizer in your car, your jacket pocket and your purse. Use them right away when you break any of the above rules and before you eat anything.

10. Do not put stuff in your mouth

11. Tissue papers or snot cloths though serve well to "wipe" the runny nose they inevitably contaminate your hands and then you contaminate everything you touch. A better product would be which has an impermeable membrane on

one side so your hands stay clean. One good example is female sanitary napkins. They are excellent to wipe runny nose in my experience as they keep the hands and environment clean. They nicely fold after use and could be used again . They are also less abrasive then most tissue papers. I seldom get "cold" or "flue" but when I do I use sanitary napkins to wipe my nose. It increases the half life of my clean hands and limits the spread of infection to the environment.

Now question is why do you wash hands at all? When no virus or bacteria can penetrate the intact skin of your hands then why wash them? Well reason is because hands are used by these germs as transport vehicles to get to area where they can actually penetrate from. And what are those vulnerable areas? Your mucous membranes! What are mucous membranes? These are the "pinkish" layers which cover the inside of all your openings or orifices such as mouth, nose, eyes, rectum, vagina and penis. All the "holes" are layered by mucous membranes which are vulnerable to viruses and bacteria. You keep your hands clean so you do not transfer the germs to your mucous membranes which you inevitably will if your hands are contaminated. You touch your faces very often, you snack very often, you put your fingers in your mouth very often, and you rub or scratch your noses very often. So having clean hands only reduces the chances of transferring these infections to your mucous membranes. So more and longer you keep your hands clean, less chances you will get sick. But can you do anything say once you have transferred the germs to your mouth, or nose or eyes etc as good chances are that you probably do it many times a day. Well there is nothing which prevents you from washing or rinsing your mouth, nose and eyes every time you go and wash hands. I as a habit always rinse out my mouth, my nose, and my eyes and wash my face whenever I wash my hands. This is not the current recommendation by doctors but it only makes sense. When doctors get splashed in the eye with a patient's body fluids they know that rinsing the eyes decreases the chance of infection. So why not rinse eyes, mouth and nose as routine. Virus and bacteria likely do not penetrate the mucous membranes right away. Wash them out when you get a chance and particularly before you eat or leave your work place. Following are the timings I recommend that everyone should rinse their hands, mouth, nose, eyes and face.

1. First thing after waking up: This ensures that you get a clean start and if you are sick then you "debulk" the infectious material so you don't transfer it to others.

2. Between 12:00 - 1:00 Pm. In other words before lunch

time. You don't want to swallow any viruses or bacteria which might have found their way into your mouth, nose or eyes.

3. Around 4-5 pm: Again making sure that you clean any germs before getting in your car and then bringing those germs home. If you don't get to do this at work then you should do this first thing after you get home along with changing out of your dirty work clothes. Work clothes are filthy no matter how good they look.

4. Before supper say around 6 -7 pm. Again making sure you get rid of any germs you might have picked up from your kids as schools are very dirty and hub of infections. Any parents of school going kids will tell you how often they get sick

5. Before going to bed say at 10 pm.

This ritual might seem excessive but you will get used to it quickly. Also washing with soap and water is the cheapest and safest intervention without using any harmful chemicals or antibiotics etc

There would be another added advantage if everyone followed these timings. Then all humans will act like one big organism getting rid of infections from the society.

Because it defeats the purpose say if you washed your hands but I didn't and then we shook hands! So if all humans wash their hands and rinse their mucous membranes at the same pre-defined times then infections have less chance of sustaining themselves in human population.

So what I am recommending is a timed "prescription" hand/mucous membrane wash, 5 times a day as listed above for all human beings to fight together against infections.

Now let's talk about the blood-borne viral infections which tend to be chronic and more serious in many ways. These infections have the ability to contaminate human gene pool forever as they get transferred from mother to the fetus via vertical transmission. Three very common such example humans are fighting are Hepatitis B, C and HIV. These infections have caused considerable suffering and deaths and continue to spread. For example third world countries like Pakistan are badly and I mean badly losing their battle against Hepatitis B and C. In certain populations the prevalence rate is more than 90%. Reason for this is wide spread practice of re-using syringes by almost all quacks which provide healthcare to majority of the public in rural and small urban areas. Poorly sterilized surgical equipment is also a major culprit. Re-packing and then selling of the syringes though illegal but has been reported frequently.

And then there are hairdressers who use the same blades on everyone for shaving etc, dentist which still use the glass reuseable syringes. And then there are fomites such as sharp kite threads which are used in fighting kites and lead inevitably to cuts on the fingers and blood from those cuts contaminates the threads which will inevitably be used by someone else and hence can easily transfer blood-borne infections. And then there are of course tattoos and piercings done to decorate your bodies which apparently need decoration. Humans are losing this battle. Third world specially is not equipped to fight or even stop these infections. Hepatitis B and HIV remains incurable and though Hepatitis C treatment has shown to be curative to some degree, the treatment protocol is expensive, lengthy and is similar to cancer treatment than say pneumonia treatment and is simply out of reach for the masses which desperately need it.

I think first world again needs to help the third world here as simply by itself third world has no chance of winning this war in my opinion, after all there is no third or first world its only one world. So what should the world do? Here are my recommendations:

1. Stop getting tattoos and piercing. I agree they are cool and if done with sterile needle and ink there is little risk of

infections. But not everyone in the world is fortunate enough to get clean needles and ink and many young people have acquired incurable infections in order to look a little cooler. Humans already look cool enough in my opinion and don't need ink. First world needs to step up and as I mentioned above boycott a practice which is leading to human suffering. Once a girl gets Hepatitis B, not only she has acquired an incurable disease which will cause cirrhosis and liver cancer ultimately, all her unborn children and their children and their children and so on will also have this disease. Many of them will not make it to adulthood and only very few might have a normal life. So see how it works. As the modern world and particularly celebrities get to be the role models for the world this entitlement comes with a very serious responsibility. And if you think your tattoo place is practicing strict sterility guideline then think again. Just last month a tattoo parlor in Calgary, AB, Canada was raided and was found guilty of spreading Hepatitis C to many of its clients. I don't want to take any chances with my gene pool and I don't think anyone should. I feel the same about piercings. Its mutilation and people who gladly go and get their daughter's ears pierced are guilty of child abuse in my opinion. So stop getting tattoos and cover it up if you already have one. Do not get any piercing and remove all your rings forever in order to show your support for fight against Hepatitis B, C and HIV.

2. Syringes use outside the main hospital particularly in third world should be banned all together. Most medications are available in oral forms.

3. Fighting kites should be banned as those sharp threads act like any other sharps and can transfer blood-borne infections. You are probably wondering how this can happen. Fighting kites use sharp threads (starched with abrasive material such ground glass) which act like wire saw and cut the other kites threads. Almost always these threads also inflict minor cuts on the hands of the kite flyers so often these threads are coated with human blood and plasma. Just like used needles when these threads are used by someone else they can easily transfer blood borne infections such as HIV, Hep B and Hep C. Cut kites can travel many miles air borne before being captured by someone else and doing this they have made these blood borne infections air borne literally.

4. Flanders "poppies" have sharp pins and these poppies inadvertently end up in hospitals and schools and in the hands of children and can transfer infections. The chance of getting blood-borne infections though low is there and truly depends on the prevalence of these diseases in the society. I am sure there are other safer ways to raise awareness and dollars for veterans. You don't need to put

your kids at risk for infections and traumatic needle stick injuries. Such an injury to the eye will lead to traumatic cataract and blindness. Its not worth it.

5. International pressure needs to be applied to third world countries to better regulate reuse of syringes, surgical and dental equipment and shaving blades etc

6. Safe sex practices don't need any further emphasis but world is putting its guard down and STI including HIV are on the rise again. Always wear protection unless you are trying to conceive and always get a parenting contract before engaging in a sexual intercourse.

7. Again third World is in dire need of "infectious disease centers" as infections such Hepatitis B, C, HIV, TB, food and water borne diarrheal illnesses, malaria, dengue, intestinal worms, scabies etc remain the major reason for human suffering, death and loss of healthy genes.

Clothing:

Humans are unique in many ways. One of the most obvious is the lack of hair or wool on their bodies. Humans are truly naked. Not many animals are designed that way. What is the advantage? Well most significant advantage is heat dissipation. As wool or hair protect against the cold, same wool becomes a problem when the body needs to dissipate or loose heat. For example cheetah when running gets over heated very quickly and then has to stop when the body temperature goes above a certain degrees. Humans on the other hand have no such wool and have abundant sweat glands and hence do not get overheated easily. In fact humans have the ability to out run any other land animal over long distance mainly because of this heat dissipation ability. This is a clear survival advantage. Naked skin also is cosmetically appealing and nature does pay attention to aesthetics y and being sexually more attractive also has a survival advantage. However this naked skin has certain disadvantages as well. For example lack of wool leaves the skin vulnerable to elements such as cold, or sunburns, dryness, bruises, thorns and gets dirty easily. Also it has no defense against blood sucking parasites such as ticks, mosquitoes and leeches. Naked skin being sexually attractive sometime leads to unwanted attention and contact by others and this problem is worse for children and women. So covering the skin as much as possible only helps to guard and protect it. You can experiment this yourself. In summer wear shorts and tees and go out and count the number of mosquito bites per hour. Then put on pants, socks and a full sleeve shirt and then count the number of mosquito bites. You will be amazed on the results. Children specially are more vulnerable to mosquito bites as they cannot defend themselves much. Mosquitoes pick their victim and site carefully or they get killed. So they target children and areas which are hard to see for example back of the legs, ankles etc. As mosquitoes kill millions of humans every year its only wise to protect your otherwise beautiful but vulnerable skins from these pests causing malaria, dengue, encephalitis, west Nile virus and many other life threatening illnesses. Again first world has a responsibility and a role to play as the trend setter. My appeal to fashion designer is to think of functionality and practicality first and then looks. It's only wise to do so. Wearing full clothes also protects against unwanted attention and touch and again it is more important for children. It's estimated that about 25% of girls get sexually abused. Why leave this on chance or to someone else's self control. You know there are evil people out there so protect your children and yourself against harm. Covered skin is clean and protected and exposed skin is dirty and at risk. And as your children will wear what you wear, it makes sense to dress for comfort, functionality and skin protection rather than just seeking as much attention as possible.

Toilet issues:

There are two types of toilets being used in the World right now squatting toilets and flush bowl toilets. Squatting toilets have one clear advantage over modern toilets and that is no part of the skin touches anything in the toilet. You do have to take your pants off completely and squatting could be difficult if you are not used to it but they keep your body clean. Squatting also is more anatomically correct way to empty the bladder and bowls. Squatting toilets could use some design improvements against splashing for example but mostly they work well. Problem with modern flush bowl toilets is that it's impossible to keep them clean in public places especially toilets for men. Most men will stand and urinate on the flush bowls. This invariably leads to microscopic splashing of the urine and water in the bowl to everywhere in the bathroom including your clothes, wall, toothbrush and the toilet seat itself. And then you are supposed to sit on those very seats with your bare skin. You should know that someone else's urine is all over your skin now. And not just any one person urine, a whole lot of people urine is on your skin. Go to any public toilet for men and you will see what I mean. Seats are disgusting. Most men try not to sit in the public toilets unless they really have to but then you do have to occasionally. So you close your eyes, wipe the seat with tissue paper and then sit on it as you have no choice. Even though you clean it as much as possible you know that you are sitting on someone else's urine for sure. Just because you can't see it doesn't mean it's not there. So what's the answer? Obviously you can't install squatting toilets, that's just not practical and will not work. Only and simple option is that everyone sits down all the time even for passing urine. Men should do what women already do. Sit down first and then do your business. This is where we need to bring equality. There is no shame in being cleaner. Do not stand and urinate on a seat you or someone else will be sitting down on soon. Sitting down prevents urine from splashing all over the toilet seat, the walls and on your clothes etc. If you don't believe me just search the internet for urine splash back and you will find your confirmation or you can ask any forensic expert to come to your house and show you how far your urine splashes on your toilet walls when you stand and void. If you can all follow this simple advice that's "please have a seat" then imagine how much cleaner bathrooms will be everywhere. Again no matter how close a practice or ritual is to your heart once you realize there is better and cleaner way of doing it then you should change your ways. That is the only way you can continue to evolve in a better direction.

If you really want to stand and void then do it in the urinals or in the bush or on the fire hydrant or anywhere else you want but not on the seat you sit on.

Third World countries like Pakistan really need to educate their masses about the proper toilet etiquettes. First of all there are simply not enough toilets for everyone there. Even if you can find toilets they are dirty and stink like anything. Part of the reason is that to build and maintain a flushing toilet is an expensive project and requires a lot of water and a sewerage system which is not always available. An easy and cheap alternative is a composting toilet. A composting toilet uses a bucket or drum to hold the excreta. And after you are done with your business you cover your excreta with either wood dust or dirt which leads to aerobic decomposition of the organic matter and turns human excreta into useful fertilizer. As there is aerobic decomposition compared to anaerobic decomposition these toilets do not stink much if any at all. These are easy to build, maintain and are portable.

Intestinal Worms

Humans are very prone to intestinal worms. It's estimated that 40 percent of humans will have intestinal worm infestations at some point in their lives. Many intestinal worms typically penetrate through the barefoot skin to get inside the body. After entering the blood stream they travel inside your veins till they reach the lungs. Once inside the lungs they come out of the alveoli and crawl upwards till the reach the throat and from there they go down the esophagus and into the stomach and from there to the intestine where they mature and lay eggs. Those eggs come out with feces and from those eggs microscopic worms come out which will penetrate someone else's foot to enter their body and this cycle continues. Other worms come out of the rectum at night to lay eggs around the anal opening. These eggs cause itching and which makes people scratch themselves. When they scratch, worm eggs get under their nail folds and then as those hands go to the mouth they deliver the eggs there. Once inside the mouth eggs are swallowed and that is how some worm maintains themselves in human population. Yet others find an intermediary animal host and humans get them by eating undercooked meat of those animals. Pretty amazing stuff isn't it. But worms are nasty and they rob a body of nutrients and vitamins and they don't always stay in the intestine and sometime can travel to other organs and can cause serious damage or even death.

How can you prevent worm infestation? In order to be successful you have to understand their cycle of life so you can break that cycle. For example freezing meat for few days and then cooking it thoroughly destroys worms. Frequent hand washing and clipping of the nails short deceases the chance of transporting worm eggs back to your mouth. Access to clean toilets is essential to prevent contamination of the soil with worms and their eggs. And always wear shoes. Shoes could be the single most useful intervention in fight against intestinal worms.

However in many parts of the world humans particularly kids cannot afford shoes. Human feet without the shoes are very vulnerable to worms, elements, thorns and injury. One easy solution to this problem could be socks with an impermeable but flexible coating on the bottom. I saw a runner who was running in such socks. He told me how he had made these by first cutting out an impression of his feet out of cardboard, inserting the cardboard inside the socks and then applying glue on the bottom of the socks. Once the glue (contact cement, shoe goo, spray-on adhesive etc all work) dries up, you remove the cardboard from inside and now you have custom made socks you can wear like shoes. I wasted no time in making a pair for myself and for me these socks are best running shoes ever.

I have run in them in below freezing temperature where my hands and face gets cold but my feet are warm from the heat being generated from running. These socks could be the solution for the kids who cannot afford shoes or will not wear shoes. These socks will protect them from thorns, dirt, heat and most importantly intestinal worms. These socks could be either donated or simply made by locals and could help the local economy as well. They are like rubber coated gardening gloves which protect your hands from the thorns.

anyone not even to themselves. People already know what to do for example eat healthy, exercise, don't drink etc but their body which has a primitive mind of its own is just not tamed enough. There body is like an untamed dog who just doesn't want to listen to its owner. So First thing one has to do is to get a control on one's body. Make sure you can get your body to do whatever you want it to do or whatever needs to be done. Start slow and don't give up. This is also important because the next 5 tips need you to be in control of your body or you will not be able to follow them. If you let your body act on its primitive animalistic instincts then it will eat as much as possible and will not want to work out and you will never get healthy. I or no one else or any

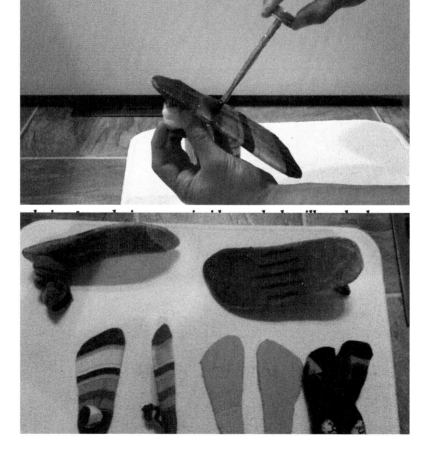

Six rules to healthy and long life

Lot has been written about how to live healthy and longer and everyone seems to have an opinion about it. Yet humans are if anything getting unhealthier over all and this generation will have a shorter life span then the previous generation for that reason. I have come up with 6 simple rules to keep humans healthy, young and fit for longer. And they are quite simple to follow and understand. It is more like common sense maintenance than anything else. Human body is an amazing machine and is well built. It does have flaws but they are mostly in the software and hardware is pretty robust. As no matter how expensive your car is you do have to do the regular maintenance on it or you will significantly shorten the life of your engine. Same way human body needs maintenance. So here the 6 essential things one must do if one wants to live long and healthy.

1. *Be the boss of your own body*:

That is the first advice I will give you. There are many studies which have shown that patients do not listen much to doctors and in fact many advocate against even talking to the patients as its considered waste of doctors billable time. I have realized why that is. Reason why people don't listen to their doctors in my opinion is simply because their body doesn't listen to can has about 160 Cal for example. Anyways back to why overloading is bad for you. Best way to explain is to think that fat storage in the body is limited. After a certain point the storage runs out and then as body has no way of getting rid of fat, this surplus fat starts accumulating in the internal organs such as liver etc. This is why fatty liver (non-alcoholic) is a fast growing reason for permanent liver damage requiring liver transplant. It's almost becoming an epidemic in western World. Other organs such heart and arteries also start accumulating fat. Fat stores in the neck make the neck thicker and start making the airway narrow. As the airway gets narrowed then the air has a harder time to get to lungs. This problem can lead to no air entry when patients are asleep causing a common condition called Obstructive sleep apnea (OSA).which means patients are unable to get air at night and the body is gasping for oxygen most of the night. They feel tired and not rested in the morning and this leads to serious more serious medical problems. This is treated by a machine called CPAP which forces air down the lungs and one has to wear it every night.

At some point body decides that it's not going to store any more calories and then everything you eat simply stays in your blood. Blood sugar level remains high even after two hours of eating and even after overnight fast. This is called diabetes mellitus. Fat levels also become high and this is known as "high cholesterol" or "dyslipidemia". Both Diabetes and dyslipidemia lead to a number of other diseases such as heart attacks, stroke, kidney failure requiring dialysis, peripheral arterial disease, erectile dysfunction, neuropathy etc. There is no real treatment for diabetes available so far. Available treatment options mainly lower the glucose level temporarily in the blood but they do not change the outcome or modify the disease course much. Insulin for examples simply pushes glucose into already full stores and hence lowers the blood glucose levels but does not reduce the rate of early heart attacks much. Only way you can really help the situation is by emptying out the energy stores significantly. But as the pancreas which produces insulin gets burnt out, it might take a lot of weight loss to normalize your sugar levels again.

Too much fat also adds to the weight and that weight has to be carried by the spine, hip, knees, ankles and feet. All of these joints will feel this extra weight and will wear down faster. Improper shoe wear as described above adds to this damage.

In short overloading your body damages every system of your body severely. In reality almost all organs and systems in the body could be damaged ultimately by overloading. It decreases your functionality, reduces your lung capacity, compromises your airway causing sleep apnea and creates complex metabolic problems such as diabetes and dyslipidemia. Another under looked problem is its ability to interfere with heat dissipation as fat acts like insulation. This reduced ability to dissipate heat leads to overheating quickly which further reduces one's exercise ability. Once you have managed to overload it, it's not easy to offload either as you have to burn and breathe out every single gram of fat in the form of CO_2.

It's actually relatively hard to put on weight. For example to put on 1 kg of fat you have to have a positive balance of 9000 Kcal or 1 kg of butter. Same way in order to lose 1 kg, you have to burn 9000 calories but of course it's harder to burn then consume. When you run 10 K, you burn about only 500 Calories. So to burn 1 kg of fat you have to run 180 km. It's not easy to run 180 Km as you probably know and even if you did body will burn mostly glucose and not fat. Restricting your caloric intake is extremely important. If you can manage to be negative around 400 Cal a day then you could lose a pound every 10 days and that would be 3 pounds per month and 36 pounds per year. This is not easy by any means though. To be negative 400 Cal you will have to restrict your intake by staying away from caloric rich foods and increase your burn rate by increasing your activity and/or exposing yourself to cold. Why cold exposure? It's because there are only two ways you can increase your caloric burn rate one is by increasing activity and two by cold exposure. When you expose yourself to cold, your body needs to re-warm itself to 37 C, it's uncomfortable but one can burn a lot of calories over short time. For many people increasing the activity is simply not an option for example if one is too overweight to exercise or patients with severe arthritis etc so for them cold exposure is the only option to increase the caloric burn rate and improve their muscle mass. A cooling blanket at night for example could be a great way to burn calories when you are sleeping. You will have to get used to it slowly, my advice would be half to one degree drop in the temperature at a time and then giving yourself few days to get used to it. This is a very under looked way of increasing caloric burn without doing anything at all.

1. Keep the body in a running condition:

Body is an amazing machine. Think of it as a car which adapts to whatever your needs are. If you like it to be all-train, heavy duty, powerful truck, it will become one and if all you use it for is getting to your car and back them it will downsize the engine as its simply more fuel efficient that way. It will give you what you use it for. So to keep it in a running condition you have to simply keep it running regularly. Either exercise or pick a hobby or a job which requires some physical activity. If you let it sit for too long it will downsize everything, its fuel storage capacity, its lung capacity, its heart capacity, muscles, bone structure etc. And then you will run into problems such as diabetes earlier. Exercise keeps you younger. Believe it or not a 65 year old human has the same running speed as a 19 year old. So do not blame age for your aging, blame yourself for your aging. Exercise is the cheapest and best cosmetic intervention out there hands down. You don't only look young, you in fact are young. Any exercise is better than no exercise. You can start with just 1 minute a day and increase it by minute every week. By the end of the year you will be working out for 52 minutes if you keep at it. Don't give up and get discouraged. Do not compete with anyone else. Every time you work out know that you are in better condition now then you were yesterday. Getting into good shape is a long process and requires hard work but it pays back in more ways than you can imagine.

1. Keep it clean:

Many illnesses simply happen because of poor hygiene as I have explained in detail above. So keep your body covered and clean. Keep your clothes, environment, bathrooms and kitchen clean and you will get sick less. Practice safe sex by wearing a condom unless in a steady and reliable relationship. Stay away from piercings, tattoos or anything which can compromise your skin as it's the best defense you have against infections. Keep your feet and socks clean and always wear shoes. Barefoot running movement though is right in a way that it makes a human run they way its supposed to and does not change the standing angles but it leaves feet vulnerable to injuries, infections and worm infestation. So wear shoes but make sure your shoes do not change the anatomy or angles of your feet.

1. Do not poison it:

Cigarettes and alcohol are the two most common ways people

damage their bodies. I have talked about alcohol in detail before so I will not talk about it here. Smoking everyone knows is bad. However I want to clarify something. It's not the nicotine which causes the damage. It's the "smoke" which is bad. Lungs are not built to handle smoke. In fact all animals run away when they see smoke or fire, humans are the only ones which will go and inhale the smoke into the lungs. Smoke simply cooks the meat for example smoked salmon. So the body is getting cooked from inside out when you smoke anything, it doesn't have to be tobacco. Marijuana smoke also has the same damage. And taking nicotine or THC (marijuana chemical) by means other than smoking does not have the same damage to the body.

Other common poisons are cocaine, heroin, PCP, ecstasy, etc. The list is long but stay away from all of them. Cocaine is especially bad as it can cause heart attacks in otherwise young people. Also cocaine is typically adulterated with other drugs such levimazole. Levimazole has its own problems and can cause death as well.

Mode of delivery of a drug can also make a drug more dangerous than it otherwise is. One of the worst things you can do to yourself is to use needles to inject drugs such as cocaine or heroin. You will breach the skin and introduce bacteria which will infect your heart and damage the valves; you can also get Hep B, C and HIV that way. So do not ever use needles.

1. Do not alter the design:

Human body is made nearly perfect in its hardware. And changing the design even slightly could have complex long-term problems. Most obvious example is shoes with high or elevated heels which add a heel-to-toe gradient. This in my opinion has been a crime done against humanity and many people have suffered and died because of it. Other example is shoes which are pointy as they force the big toe to point outwards in an unnatural way. Stick with zero gradient shoes which complement the foot and not modify it and wear clothes which allow normal range of motion of all your joints in the body.

If you follow above simple guideline that you will be able to prevent most of the common diseases humans suffer from such as diabetes, hypertension, arthritis, COPD, lung cancer, OSA, fatty liver disease, alcoholic liver disease and other alcohol related disorders, heart attacks, stroke and many infections. If you fail to follow any of the above recommendations then know you will sooner or later pay the price. And if you already have any of the above diseases then you most likely have violated one or more of the above rules.

Body is a like a tree, you cannot change it in one day. However if you take care of it a little bit every day then you will have a healthy tree and if you ignore it then you will have a sick tree and it will take a while to get it healthy again.

Evolution:

Evolution is considered a scientific truth now. Humans evolved from early Hominids which evolved from apes which evolved from other primates and hence we can trace this to all the way to pica which were fish like creatures and were the first to have a "spine". However to me it is unclear how the evolutionary biologist have concluded that whether it was passive or random evolution or was it a design upgrade or "active" evolution. For example if you look at a cell phone from 1980s and today's Iphone 6, you can clearly establish the evolution here. They both are built very similarly; they have a chip, screen, battery, antenna etc. But you know this did not just happen randomly. So as life forms evolve how can you not entertain the possibility of active or supervised evolution?

By definition evolution can never ever achieve perfection as it doesn't know what perfection is. It's just random mutations and if it works it survives. As a continuum from simple to perfection it starts from nothing and is randomly finding its way towards perfection. But when one looks at any life form and how it integrates into its environment you cannot help notice perfection. In fact perfection is the theme of life. You see a fish and realize how perfect its design is, streamlined, able to extract oxygen from water, find food, reproduce, camouflage and move with such elegance and perfection. You can barely see a Gabon viper hiding perfectly among the fallen leaves. A viper did not need to be so perfectly camouflaged to survive and there are lots of other vipers surviving without having the same elaborate markings. And there must have been other similar patterns too then and they should have survived as well. Only where there are leaves like that there are vipers like that! Same way when a bird is flying, think of it as an airplane flying as the bird is following all the laws of physics. A common house sparrow flight from one branch to another is an amazing thing to watch from engineering point of view. That how quickly it can take off, maneuver itself in air and land perfectly on a twig. It's nothing but perfection and scientist can't even imagine improving upon that design. What about humans. A biped weighing 160 lbs, walking perfectly balanced on a rope against gravity, wind and rope oscillations. Ask any physicist and she will tell you that this is pure perfection. You cannot even think of designing a better machine. This is absolute balance. Homo sapiens did not really need to walk on ropes stretched across canyons to outlive the other Homos. Then why and how randomness achieved such perfect and well calibrated hardware. I think this degree of perfection in designs, camouflage, attention to details for sake of making it look pretty and reproduction ability can only be achieved by something which knows what absolute perfection is and how to achieve it. This obvious "perfection" in design and execution of creation of life is a valid argument in favor of active evolution in contrast to passive or random evolution. Passive evolution is a pretty naive concept in my opinion. It's just like saying, oh it's too complicated, too sophisticated of a design, we don't understand how it works so as we don't understand it, it must have happened just randomly by its own over time. Iron molecules randomly found their way into blood and happen to have the ability to bind oxygen reversibly and transport it to where it was needed and all these chemical reactions randomly happen at the absolute ideal oxygen partial pressures. Binding of oxygen to iron and then releasing of the oxygen in the tissue is a clear evidence of absolute chemical perfection as all those chemical reactions happen at the ideal partial pressures of oxygen. It's not possible to improve on the design. Someone or something knows exactly how everything works. There is no doubt in my mind that life is created with attention to every little detail in order to achieve perfection.

Animal Cruelty

Why do humans still need to harvest fur? I understand in good old days fur was irreplaceable as it was the warmest material available to humans who lacked fur of their own. But now it's different. Synthetic materials or wools are available and can even outperform fur. Then why still trap helpless animals for just their furs? Why trick them and torture them and kill them simply to sell their skins. It's a mindless cruelty and that's why fashion for sake of vanity alone is disgusting as it leads to many innocents lives getting destroyed. Fur harvesting is cruel as a hobby and as an industry. Anyone wearing fur for a fashion has paid someone to torture an animal and then kill that animal to take that skin so one can wear it and look good. Inside a fur wearing fashionable person is an ugly creature no matter how good on the surface that person looks. There is no reason for this. Leave the animals alone. You did not create life and you don't have a right to take it. Treat animals with respect and give them the freedom they need. I have seen fur harvesting videos coming out China where animals are being skinned alive. This was one of the worst things ever to witness in my opinion. These are the moments I really cherish the fact that I know God is there so I know no one is getting away. If you do need to take life of an animal for food then please make sure that animal when alive lived a happy life and then was killed in a human and quick way. Taking permission from God who created life makes sense to me as it's only courteous. Life should not be taken from some living creature just for anything. It should have a purpose. And only be done for your own survival. So killing for food, pest control such as rats etc makes sense. Killing for harvesting fur for fashion does not. Also killing for fun or target practice is mindless.

Same way hunting ducks though fun and provides food is very cruel. Ducks come in tight flocks and when shotguns are fired pellets don't just hit one duck, they may hit one fatally but they almost invariably hit other ducks as well which though manage to fly away must suffer the pain while on their long biological journey. No one knows what happens to them but I am sure they die a painful death from starvation or infection.

I want to say a lot about animal cruelty as my heart cries the most when I see helpless and speechless animals suffer unnecessarily. What has come over humans? In order to make money you are killing everything slowly but surely. All life forms are products of DNA just like us. We need to preserve as much as possible the genetic data of all life and not just Humans. If humans are not going to take care of other life forms then who will?

Please stop collecting preserved insects, butterflies, birds, fish or any other animal as a hobby. This industry is destroying life from the planet at a very fast rate. If you really want a hobby then learn how to tie flies and make your own insects from threads and wings etc. Don't just kill a butterfly so you can hang it on your wall.

My message to all Governments head is that know that all life form under your rule is your responsibility. A lot more needs to be done then currently being done to protect the animals. I salute organizations such as PETA trying to defend animal rights.

Pesticides:

Pesticides are poisonous chemicals used in agriculture to kill insects or other pests such as rodents. However my problem with pesticides is how they interfere with food chain. In nature nothing goes to waste. A dead insect becomes food for a bird for example. However when pesticides are used this rule gets violated. A poisoned dead insect will kill the bird that eats it and that dead bird may kill the coyote who eats it. This problem is wrecking havoc in the animal kingdom. Yes pesticides have increased production of the crops but at what cost to the planet? These pesticides then leach down into the soil and water and contaminate subsoil water.

In this day and age humans have the technology to grow crops without the pesticides. Such as perma-culture where up to 7 different crops are obtained from a single piece of land without the need for any pesticides or fertilizers. Common example is very successful "bird friendly coffee" also known as "shade grown coffee". This method of farming is not just viable but more lucrative, more environmentally friendly and simply more humane. Humans need to learn to live together with other animals. Other example is cattle farming done in Alberta, Canada. Cattle are allowed to roam in the forest for grazing. This keeps the forests intact allowing for the biodiversity, lumber industry, hunting, camping, hiking and other activities at the same time. Other example is "grass farming" aka "management intensive farming". There are many ways to grow food without poisoning the food chain.

Housing Schemes:

One of the best athletes I ever met is my good friend RA. He still holds our college 5000m record. Reason I think he is the best athlete is that he never got any training at all and yet he was the best runner, cricket player, football player, handball player, badminton player and anything else he played. He would throw a stone and literally you will watch in awe as you could barely see the stone land. It seemed to continue to keep going. He had in him and still does a World class athlete. What was his secret? Obviously genetics played a role but there was another factor which played a role. He grew up in a tiny one room house with 7 siblings. So what RA lacked in his house was space but just outside his house there were grounds everywhere. He was always outside playing with something. He was either kicking a ball, throwing stones or running etc. Secret to his success was small house big grounds. I think this is the way to go. Developers should first build a ground and all around it put small houses so kids can just safely play in a common ground without having to be driven somewhere. Housing societies should be centered on grounds and not the other way round. It will be also cheaper to maintain these common grounds. A ground doesn't have to be very big, just big enough for kids to play.

Price of peace

Peace is in short supply these days. Wherever you turn you see unrest, protests, displaced people and violence. Organized protests, wars and pre-meditated crimes have always been around but what worries me the most is rise in mindless crimes where someone loses control, snaps and ends up killing good people for no good reason. For example a black guy shot two police officers in cold blood in New York for no personal reasons. It was in response to how police officers are treating black people across the country. So the man was angry and his emotions took control over him. A 15 year old kid was intentionally run over and killed by someone in Missouri. Why? It wasn't personal it was just because that kid was Muslim and the killer had apparently hated Muslims so much that he could not control himself. Again he was simply being controlled by his emotions without thinking at all. Few days later a Muslim kills two people in Australia. Then 12 people get killed in France for making cartoon about Muhammad by two Muslim men. Within a week of this incident about 50 different hate crimes got reported against Muslims all over France. There are other similar examples of this ping pong violence which seems to be gaining momentum. What appears to be the theme behind all these crimes is increasing racial and religious divide in the World.

History tells very horrible stories when such mindless violence erupts. For example in 1947 during Indo-Pak divisions millions of humans got killed as a result of this ping pong violence by both Hindus and Muslims. You think humans are any different now? They are not; if they were there would be no such events happening. Humans are highly prone to getting sucked in by this kind of hatred. Hate crimes are like fire and hate only leads to more hate and this become a self-sustaining process just like an uncontrolled nuclear reaction. Mob frenzy takes over and then people go around killing other people just like zombies. This has happened many times such as Rwandan genocide not too long ago and is still going on in certain parts of Africa and Middle East. And once this starts there is no way to stop it as people lose all their abilities to reason without realizing they have lost it and law and order is usually already non-existent by then. So there is no means to intervene at that point. It only stops when there is nothing left to destroy. Just like a wild fire going out of control, it only stops when it runs out of things to burn.

But there are ways to prevent it. These isolated incidents I mentioned above are like sparks of fire. If there is enough dry wood around then they can start a fire quickly. And these sparks are a clear sign that there is plenty of "dry wood" and it is just the matter of time that these destructive fires will erupt. In order to prevent it one has to understand this phenomenon better and what are its pre-requisites. First pre-requisite is dry conditions that is people are desperate and fed up already, there is not enough food or jobs or there is perceived injustice, oppression, discrimination or threat etc. Second is division of the society into groups such as for example it was Hindu vs Muslims in 1947 Indo-Pak division, Christian vs Jews in holocaust and Hutu vs Tutsi in Rwandan genocide. One group perceives themselves as right and civil and the other group typically a minority is considered bad or evil and is seen as a problem. That is why Nazi came up with "final solution" as they saw Jews as a problem needed to be solved. It is the self assured "civil" group which causes the destruction and is basically acting anything but civil but does not perceive it such in the heat of the moment. They are 100% sure that what they are doing is absolutely right and they are doing a great service to their country or religion. At that point they consider killing innocent people as their moral or religious duty. This is exactly what ISIS is doing right now in fact and that is why no one can talk to them as they have lost all kind of reasoning. But what ISIS is doing in Syria and Iraq and is not going unnoticed by the rest of the World and this is creating an emotional response in people watching their mindless actions. But what people do not realize is that this emotional response, this anger and hatred towards ISIS will turn them into people like ISIS and soon they will start killing innocent people around them just to get even with ISIS. That is only going to make ISIS more aggressive and do more mindless acts and their actions will make people in the civilized World more aggressive towards Muslims. And this ping pong violence will continue till there is no one left to kill. Already in India you can see how anti-Muslim sentiment is gaining strength and not just in India but it is happening all over the World. However it may not manifest in countries where there is justice, equality and plenty of foods to go around as people are not desperate. But in countries where there are "dry conditions" and people are already desperate due to something else they are at risk of snapping. Even in modern countries like USA the conditions are somewhat dry now. For example there is over 50% unemployment rate in the neighborhood where recent Baltimore riots happened.

In short mindless acts are like fire and they spread just like fire. But exactly how do they spread? It's the simple news of such events which makes these fires spread. This is what happened in Indo-Pak division. Few Muslims or Hindus were killed first and the news got out and as a result more people were killed to make the score even but then score never gets even in such situations and when it was all said and done 1 million people had been slaughtered. Whenever someone hears the news of such events it triggers an emotional response in that person. And when you put desperation, lawlessness and mob frenzy together then most humans cannot control themselves and turn into animals without knowing what has happened to them.

Crimes need to be solved and criminals need to be brought to justice but hate crimes when broadcasted they only lead to more hate crimes and media plays a big role in the spread of such crimes and violence. This fire is simply being spread by the news media and social media. For example those two police officers would not have died if it was not for media. The 15 year old kid would not have gotten run over if it was not for media. Media is bringing this hate and fire to everyone's heart and in this fire everyone is getting consumed. Media has to be smart, responsible and careful about it or it will end up destroying the very society it is trying to keep "informed".

Remember the days when streaking during sporting events became popular? How was it controlled? It was promoted initially by media and then later controlled by media. Media simply stopped showing it and the problem got solved. It was an act of madness and research tells you that acts of madness are infectious and simply broadcasting their news will spreads them. Another good example is suicide. Every time someone's suicide is broadcasted on TV, it leads to more suicides. Media is hiding behind "freedom of speech" but it is destroying the society structure and is leading people towards more division and unrest. Acts of madness by one person or by one group do not represent the whole religion or race and should not be shown on TV like this. Media can chose to either keep the society civil and peaceful or divide it and bring suffering and misery to it. And unfortunately it is choosing to divide it for selfish reasons. I suppose there is something to be gained from spreading this fire. For now there is still justice and people are able to keep their emotions in check as they are afraid of legal consequences but let me warn you the moment law and order breaks down then the real animals inside will come out in full force. When a human is led by hatred and has no fear then he is capable of some horrendous crimes ever witnessed by this planet. Otherwise good humans are guilty of cutting of breasts of nursing mothers! And it can happen to otherwise civil and respectable people. This is well known psychology phenomenon known as "Lucifer effect". It was brought to attention by the famous Stanford Prison Experiment conducted by Professor Philip Zimbardo. Later Zimbardo witnessed the same evil transformation of otherwise educated and civil Americans at the prison of Abu Gharib. Zimbardo wrote a book called "Lucifer Effect" which is a must read for every human in my opinion. It shows how we are all vulnerable to our circumstances. What is perceived fair and unfair by humans is based on the situation. Your ability to be good or evil is based on the circumstantial forces. If you see acts of madness on TV it can make you angry and resentful and may trigger you to become evil yourself if the law and order broke down or you somehow became anonymous for example in a mob situation. If you think you can get away with something then your evil side comes out in full force quickly. The concept that humans have the innate ability to know right from the wrong is wrong. Right and wrong is defined by the circumstances which could change very quickly.

Media has to stop showing any act of madness by a single individual or by a certain group and that includes school shootings, suicides, animal abuse cases and hate crimes based on racial and religious reasons etc. News of such events only leads to more such events and does not serve any useful purpose at all except disrupting peace. Such events should be only brought to the attention of authorities in charge of bringing justice to those people and maintain peace.

Human population growth rate:

A genome or DNA is nothing but a code or a cut key which dictates the exact sequence simple molecules line up to make a life form. Everything from shape, size, color, function of every organ and every little micro detail of that individual life from including its behavior and emotional reactions etc are precisely coded. Whether it is a tree, butterfly or a human all life forms are cut from these keys. In fact genetic codes are very similar to a computer code. Just like how every computer code is nothing but 0 and 1 repeated again and again, DNA of everything is also based on two base pairs (A-T and G-C) repeated over and over again. Hence a single strand of a DNA could be copied and then reversed copied making infinite cloning of DNA possible. It is a language but no one can understand it fully. There are about 3.2 billion base pairs which make up a human genome. A simple mutation from say A to G or a point deletion in the code sequence leads to diseases such as cystic fibrosis etc. Everyone has a unique code and if there was an easy way to put life in these cloned genomes then you can make as many of the same individuals as you like just as every time you punch in a code of say a flower on your computer the exact same flower reappears. Life forms are in a way 3D printed machines which find their own food, defend themselves and write new unique codes with which new life forms get printed and this process continues by itself. Really life is an amazing and mind blowing technology.

All life form has a very predictable life span and will die after living for a certain time on the earth. Life span varies significantly and could be as short as in minutes to as long as many centuries for certain trees for example. However everything has a life and will die off one day and the hardware will get recycled. However the genome or the code of that individual life form continues to live in the form of its offsprings. More offsprings one has, better chances that one's genes will continue to survive. If one does not leave any offspring than that is the end of the road for that particular genome and should be considered as the "genomic death" or "genocide" as that genome is no longer "alive" in evolutionary terms. Evolution is all about genomes. Consider evolution as a very long relay race where an individual passes on its genes to the next generation and this race continues. So in one form or the other you have all been "alive" for over a billion years. In order to stay in the game of evolution one has to leave behind offsprings otherwise its game over for that genome. For example childhood cancer which kills a child before she had a chance to have any offsprings is a different kind of death than an adult cancer where one has offsprings before dying. Childhood cancer not only kills the individual but also its genome so it's the "genomic death" for that genome. Loss of every genome from the gene pool shrinks the overall gene pool as each genome is unique. Old age cancer merely shortens the days one lives on earth but otherwise has no significant evolutionary impact as the genome has already been passed on and is alive.

In other words anyone not having children is basically getting kicked out of the game and that genome gets deleted permanently from the "gene pool". In computer language analogy genomes are files which contain the written code for an individual. Half the files are blue (male) and half red (female). All files are programmed to get destroyed after a certain time. Only way the "gene data bank" can maintain and grow is by creating new files by sharing half of its code with another file of different color. These new files will continue to replace the deleting files and this goes on forever. However if a file does not get to share its code with another color file before it gets deleted then that results in loss of all the unique codes in that file forever. However most files will be able to make new files and hence gene pool gets maintained. Over time codes have mutations which lead to evolution and overtime completely new species develop as per the evolutionary theory. Every file is programmed to make as many as possible new files to ensure the survival of their codes and it also increases the size of overall gene databank.

Now keeping this in mind you can understand that in evolutionary theory any adaptation which leads to less offsprings is considered "deleterious" because it decreases the chance of that gene survival in the long term. So when you look at the current sexual practices of Homo sapiens then it doesn't take long to understand that what they are doing is deleterious to their own survival. For example practicing birth control, late marriages, homosexuality and monogamy would all be considered deleterious adaptation in pure evolutionary biological terms. Over the next few generations the impact of these practices will be quite obvious. Many of these have become politically sensitive topics and this has prevented many scientists to openly discuss these issues. But I am not taking into account the current political, cultural, religious, emotional or any other factors into the account. I am just discussing these issues as a pure biologist who is studying the last remaining species of genus Homo. My goal is to preserve as much of gene pool as possible or this species goes extinct.

Homosexuality; is it natural?

Homosexuality is a practice where two members of the same sex that is two male or two females make a mating pair or have sexual intercourse .Homosexuality is an interesting phenomenon which is observed throughout the Kingdom Animalia. Many animals such as rams, monkeys, pigeons etc make strictly homosexual pairs. There was time when homosexuality was considered a psychiatric disease in humans but not anymore. Now it is considered "natural" as it exists in nature or a "variation in norm". This mating however does not result in fertilization of the eggs which is a prerequisite for creating new genomes so no offspring comes out of this practice. And as sex or mating means something completely else to a biologist than a lay man hence I cannot call this mating a "variation in norm". To be called a "variation in norm" it would have to deliver the same results that is create new life as the normal heterosexual mating does. Homosexuality doesn't deliver that. Strict homosexuality leads to no offsprings so by definition it's a "deleterious adaptation". And as discussed above no offsprings means "ultimate genomic death" and game over for the genome so it's more like a childhood cancer in pure evolutionary terms. For example if two male pigeons became homosexual than though they live a normal life span but their mating will not result in any new life and when they die it will be a "genomic death" as if those genes never lived.

In computer analogy when two same color files exchange their codes it does not result in creation of new files and as the original files will expire with time they get deleted from the gene data bank permanently leading to decrease in the size of gene databank. So if exchange of codes between the two different color files leads to new data being created over time and increase in the size of gene databank, exchange of codes between two same color files leads to data being erased over time and decrease in the size of the gene databank.

Now if you look at this practice from animal physiology point of view then again there is nothing physiological about it. Sperms for example serve no physiological role inside a rectum. Anal sphincter is not designed to be penetrated retrograde. In fact this practice interferes with the intended purpose of the anal sphincter by causing traumatic injuries and infections. Same sex mating does not serve any physiological purpose at all.

From evolution point of view homosexuality leads to genomic death. Why it still exists in nature is a separate topic and I will brush upon that later. But first let us take a hypothetical situation where 100 % of adult population becomes homosexual. This will simply result in complete wipe out of the gene pool in a single generation or in other words a complete "genocide". One can easily test this in a computer model by making all files exchange code with same color files which will result in no new files and as the files will expire all the data will be lost. So question is what percentage of population could become homosexual without posing a threat to gene pool survival or causing a complete "genocide" over time? The answer depends on the population growth rate and if the animals in question are monogamous or polygamous. If the animal population is growing fast and they are male polygamous then a higher degree of homosexuality could be tolerated in male members without any significant consequences such as rams where 10% of male rams are strictly homosexuals but as single heterosexual ram is able to fertilize many females this high prevalence of homosexuality among male rams is inconsequential. Also in male polygamous animals male homosexuality will have a lot less of an impact on growth rate then female homosexuality because a single male is able to fertilize multiple females.

But if the animals are monogamous and have a slow or negative growth rate then high prevalence rate of homosexuality will have significant negative effects on the population growth rate and will lead to decrease in the size of the gene pool. Simply put prevalence of homosexuality will have an inverse relationship with the population growth rate in monogamous animals. Higher the rate of homosexuality, lower the population growth rate if all other variables are kept the same. This could be demonstrated by a computer model. Say first 100% of files are creating just enough new files to replace the disappearing files and hence maintain a steady databank. Now if you make 20% of the files stop creating new files then with time as the old files expire the data bank will shrink by 20% every time. So prevalence of strict homosexuality in any population is the rate of naturally occurring genocide.

Prevalence of Homosexuality = Rate of Naturally occurring genocide

5% Homosexuality = 5% Natural genocide

20% Homosexuality = 20% Natural genocide

50% Homosexuality = 50% Natural genocide

100% Homosexuality = 100% Natural genocide

Even though biologically Homo sapiens are male polygamous, these days most of them at least in civilized World make monogamous pairs when they want to have children. They are also facing the problem of not only slow but in fact negative growth rate. So if the prevalence of homosexuality was to increase somehow it will obviously further decrease the population growth rate.

But question is can the prevalence of homosexuality increase or is it fixed? No one knows the answer to that question for sure. It is unclear how homosexuality maintains itself in any population. It is also not clear if it is purely congenital or acquired or both. If it is acquired then how does one acquire it is also unclear. Believe it or not no one knows its true incidence and prevalence among humans. This lack of understanding about homosexuality among humans makes me nervous. Does it or does it not spread? If it's acquirable as someone can simply choose it then obviously it can spread. If it spreads then we need to know how it spreads. Does it spread from person to person like any infection? If it is a personal choice then what are the impacts of role modeling, peer pressure and social media on its incidence? If it's a matter of choice then can it go "viral" just like a fashion does? In other words, does homosexuality have a tipping point?

Humans' sexual orientation is a complex issue. Many are surely born homosexual so let us call that condition "congenital homosexuality" but many simply adopt or chose it for various reasons for example lifestyle preferences, lack of easy availability of opposite sex, influence of role modeling etc so let us call it "acquired homosexuality". The question whether homosexuality is a choice or not does not matter as it could certainly be chosen if one wants to choose it.

The question why congenital form of homosexuality exists in nature at all has confused many biologists. First of all most of these cases are sporadic and not hereditary so its existence does not necessarily have to have any survival advantage or useful purpose. In nature diseases exist as well with certain prevalence and it does not mean those diseases for example childhood cancer has any evolutionary advantage or should be considered "natural". It is possible that homosexuality might serve a purpose of "cleaning" the gene pool of certain undesired traits or mutations. Many mutations get eradicated from the gene pool by taking away the ability of those individuals to pass on those mutations. Whenever a mutation is simply not compatible with continuation of life, it gets deleted from the gene pool. This way only the desired genes survive which maintains the health of overall gene pool. This is commonly known as "survival of the fittest". When you look at people with congenital homosexuality they do appear to have for example somewhat decreased coordination and generally one could pick up on those characteristics. It might be that those very traits are naturally being deleted from the gene pool to limit their spread. So yes homosexuality is "natural" as it exists in nature but it's a "natural genocide".

Homosexuality is being considered more and more "normal" sexual practice and many people are simply opting to choose it for variety of reasons. There are many factors involved in making these decisions such as lifestyle, role modeling, increase availability of sexual encounters and media glamorization. Homosexuality does have the potential to become more popular as it will be the path of least resistance. When all social taboos and barriers are removed and you have succeeded in completely confusing the younger generation that homosexuality and heterosexuality are essentially equal and normal choices then one can see how its prevalence could increase. It might be easier to find consenting same sex partners compared to opposite sex partners. Life style might be more "enjoyable" and "safe" given it doesn't lead to children which tend to take life on a different and not so fun path in many ways. There will be less awkwardness and gender differences in likes and dislikes etc. But there could be any other number of factors which may make homosexuality as a trend or the cool thing to do or the rebellious thing every coming out of age generation does to announce its arrival. So if for whatever reason homosexuality does tip then has anyone even thought about what the consequences of this will be on human population? Already the civilized World is struggling to maintain its population and if the prevalence of homosexuality was to increase then it will only decrease the population growth rate further. How would you even stop it if its prevalence starts getting above a certain rate?

Can homosexuality go "viral"? Of course, anything which can spread can go viral especially when you have poor understanding of how it spreads. Only time will tell for sure but this is something you would not want to happen. And once it does go viral there will be no stopping it for sure.

If the gene pool is like a big data bank then homosexuality is deleting the files just as a computer virus does. It prevents the files from multiplying by making them exchange data with the same color files and as the time passes these files ultimately expire without making any new files. It probably spreads very similar to a computer virus infecting other files when it comes in contact with. A homosexual individual can turn another individual into choosing homosexuality and this way this practice can sustain and spread in the population. It is a deleterious adaptation, a pathological practice and capable of delivering a "genomic death". Whether it's a choice or not is a separate question. If someone does choose it on purpose then it should be classified as a "genomic suicide". One should not call homosexuality a "variation in norm" as there is nothing normal about this practice or adaptation from biological point of view. It is clearly deleting the gene pool and you should not be parading about it or promoting it as it could be a very slippery slope. Best way to handle this sensitive issue is to first give the status it deserves. Do not confuse the younger generation by saying it is equal choice to heterosexuality as they are completely two opposite choices. Counseling, understanding and education should be offered to people of congenital homosexuality. It is unfortunate that their genes will no longer survive in the gene pool but such is life. By no means it should be glamorized on media or portrayed as something cool and something to be explored. Do not celebrate it in schools.

However no one should discriminate against any other human regardless of their color, race or sexual orientation. All humans are simply equally humans and have equal basic human rights and needs. Homosexuals need more understanding and compassion from the rest of the people because their genes will be exiting the race of evolution forever.

Polygamy and Homo sapiens:

Polygamy is a practice which is common among many animal species and serves an important biological role in ensuring the species survival. In humans or at least the modern civilized world this practice has somehow been given a negative social status. It's looked down upon, not accepted and is illegal. I am not sure what the basis for this law is. I suspect the law is based on Christian or Hindu religious values. Reality is Just like all other Hominids, Homo sapiens are biologically male polygamous. Polygamy makes perfect biological sense for Homo sapiens. Male Homo sapiens are significantly bigger than female and that means males compete for females and practice polygamy. Even Homo erectus and other Homos lived the same way. From evolution point of view male have many clear advantages over females. The most important of them is that male can "export" their genes and this quality gives them an ability to impregnate many females at the same time and multiply their genes many folds. As evolution is a race against time this is clearly an advantage over females. However as male and female are simply the same specie it shouldn't be looked at this way. This only ensures that the fittest and perhaps the wisest genes get to carry on. Females however cannot export their genes and can only have one child at a time so they don't need to compete or fight. For example in a lion pride a single male will have multiple lionesses living in pure harmony.

Similarly a man is also designed to have multiple female partners, its anatomy, physiology and psychology is programmed in such manner. Limiting a man to one woman by law is in fact is a clear violation of his biological rights and interferes with laws of nature and evolution. This may be playing a role in human negative growth rate and high divorce rates. For example if human rules were applied to elk, limiting one bull to one cow then that would seriously jeopardize the whole social order of elk and will threaten their very survival in the long run. Same way male lions cannot be forced to follow this rule. Biology dictates how things work and interfering with biology will have serious consequences for the very survival of the species.

For humans male polygamy is practical and physiological. A man can mate with more than one woman and in fact total societal cost of a "family unit" and raising a human child will go down if a man shares a household with multiple females living in harmony. Mother and father of all children born will be clearly identifiable and there will be no ambiguity about responsibility. There will be free childcare available; more members in a family will divide the workload etc. For example if a pride has only one male and a single female its survival is harder compared to when there are multiple lionesses. In fact many culture around the world live in polygamous families happily.

Female polygamy for Homo sapiens is not biologically viable. A human female usually has only one egg so there is no biological role for female polygamy. In animals such as dogs, females have more than one egg and perhaps multiple male partners make sense as there is always a chance that some eggs may still remain unfertilized. So females are ensuring that all eggs get fertilized and males are hoping that their sperm may still have a chance. Also as numbers are important to a pack both male and female are ensuring that all eggs are fertilized. Homo sapiens are not pack animals. A man will not tolerate ambiguity and a human child also has a right to know who his /her father is clearly. Genetic testing though could solve this problem but again we are talking biology here not equality for sake of equality. Men clearly have more sexual freedom then women biologically as they don't get pregnant which limits a woman's ability to reproduce further. Men after getting one woman pregnant can go get another one pregnant just like a bull elk or a male lion.

I want to make clear though that it is lot safer to just have one partner. And If a man does decide to have more than one partner then of course women have to agree and its man's responsibility to do justice between them and treat them equally.

Hypocracy of the current civilized society is that a man can have multiple sexual partners as long as he doesn't marry any of them! That is absurd rule and will only discourage men from getting married. Marriage is simply a "parenting contract" and it should not be tied to rigid legal rules on consenting adult human beings. Every couple should agree upon their own parenting rules before engaging in any sexual activity. Any sexual encounter with or without contraception can result in a child birth so it's only wise to have a "parenting contract" in place before hand. A parenting contract will also ensure that human beings are selecting long term mates rather than just randomly sleeping around like chimpanzees.

Divorce or breaking the "parenting contract" could be made a lot easier as well. Such difficult rules again prevent people from getting married as it's so hard to get out of it. After three months of breaking one "parenting contract" a woman should be allowed to enter another contract. Three month gap will remove any doubt that she is pregnant or not. If she was then her X-partner and the new partner have a right to know obviously. A man obviously does not need any waiting time restrictions as they do not get pregnant. Parents should expect their teenage children to have such "parenting contracts" in place before becoming sexually active. This is the only sure way to solve many of the unwanted pregnancies related issues. Parenting contract will also bring some much needed seriousness to becoming sexually active as it does have serious consequences.

Biology is sexist and gives different freedoms, roles and rights to males and females and it's only wise to learn and accept these realities no matter how hard they appear. In the end you will be happier or at least ensure the survival of your species. Refusing to accept reality and placing legal restrictions which are against biology will threaten the species survival ultimately. No one in right state of mind will restict a lion to a single lioness, then why are men being subjected to that unfair rule? This is because of certain religious practices which have been accepted as a norm without any regards to science and are being forced upon everyone. If you are not a Christian or Hindu and your beliefs allow you to have more than one willing partner at a time then there is no reason a man should be denied his biological right of polygamy by arbitrary laws.

Perhaps it would be better if the World would have consulted a biology book and not a religious book before making laws about human reproduction.

Religions and their wars

What exactly is a religion? There are different definitions of religion and it means different things to different people. To me religion means the truth about this universe and origin of life. There are lots of religions out there and then there are many sects within those religions. Seems like anyone who could imagine something has imagined something and then starts believing it to be true. Just because you believe something to be true it does not mean it is actually true. There has to be only one scientific truth about the origin or life and what happens after death and everything else is just figment of imagination and false. Lot of the religions do not make sense at all but yet people will not challenge the beliefs of their parents and will live in this make belief World that out of all the religions of the World mine is the only one which is the real scientific truth. Sometime smart people will even realize it that their religion does not make sense but they will still continue to believe in a make belief concept. I don't understand that. Really this kind of behavior does not suit logical human beings. Again there is only one scientific truth about the origin of this universe and origin of life and what happens when we die. Truth is what it is and anyone beliefs has no bearings on the actual truth. And there is only one truth. Only one of the religions is right and everything else has to be wrong. Now as smart humans our job should be to find out what is the truth and not keep walking the paths of our parents mindlessly just because we have emotional attachment to that path. What humans do is to defend that their version of the truth is the right truth just because their forefathers said so. And then humans fight about it and start wars, kill each other and try to force their version of the truths on everyone else.

Why don't we sit down and discuss the truth about the origin of the universe and life and what will happen to us when we die etc. Let's discuss like adults or smart logical human beings should discuss anything important without getting emotions and egos involved. We know there is only one true version and yes you think it's your father's version which is the ultimate truth but you should know that you are biased.

To me there is absolutely no question that life is a smart design and not just molecules randomly put together. DNA did not just get written on its own, it's written by someone or something who knows this coding language. And not just the language but it knows how every law of science works. I don't know how it is possible for something to know everything but there are a lot of things I don't know and I accept that. Atheists and polytheists are about to get a surprise of their lifetime soon. God does exist as the signature in all the created being is the same so there is only one entity creating everything. There is no doubt a "Creator" or an "Engineer" designing life. You can all it whatever you want but life is created by a Creator. Lets for argument sake call it "God" for now.

"God" theory states that there is something or someone which has 100% of the knowledge in the universe, is present everywhere in the universe and has all the power there is in the universe. Somehow God knows even if a leaf falls in the forest and even the thoughts inside your head. Question in lot of people mind is that is it even possible for "God" to exist and if possible how is it possible for something to know everything and be present everywhere and know even my thoughts. Well the answer is pretty simple. There are some proven scientific entities which can easily fulfill this description. For example "Higg's field". Its present everywhere and it dictates every atomic and subatomic particles movement. So when I think of something and it creates a detectable voltage in certain part of my brain then of course Higgs field knows it. So if Higgs field can exist so can God. Just because science has been unable to find it doesn't mean it's not there. Humans only know less than 1% of all the knowledge out there. Big deal you have travelled to space it doesn't mean you suddenly know all there is to know. But something does. It could be like a grid which encompasses and understand all the laws of all the sciences out there.

Now question is if all the Holy Books that is Old Testament (Tora), New Testament (Gospel) and the Final Testament (Quran) are messages from the same Creator. I myself absolutely believe they all are. I have tested Quran and it is true to its words and hence I now live by the Book. I was not always like this. But if you have doubt then you can conduct a controlled scientific experiment where you make one city "live by the Book" for sake of an experiment. You will see that suddenly that city will have zero crimes, no poverty, justice for all, people will be a lot more healthy, prosperous, clean and happy. There will be very little suffering and misery if any at all. Creator asks humans to be the ideal human being and overcome their flaws. But unfortunately humans are unable to overcome their flaws and end up either changing the message as in the case with Tora and Gospel or taking out an interpretation which justifies them to continue with their flawed behavior as the case with Quran. What is happening in the Muslim World is a prime example of people not following the message as it is supposed to be. Same phenomenon happened with Jews and Christians and now is happening with Muslims and Quran very precisely outlines it.

Humans are supposed to be God successive authority and were sent to Planet Earth as a short test. Anyone who acts like a human should act that is he/she understands the chain of command, is logical, honest, kind, polite and charitable will go live in the real life forever in another planet called Heaven. People who are not worthy of being called humans that is they are dishonest, cruel and do not understand the chain of command will get punished and live in a very hot place called "Hell" forever. Everything one does is being recorded and there is going to be absolute accountability for one's actions and intentions behind those actions. There is no compulsion and everyone is free to choose its path. Everyone is only accountable for their own actions unless you misguide someone and then you are responsible for their mistakes as well along with them. There is no getting away and justice will be served to all equally. All humans shall die and everything on this planet will remain on this planet which will ultimately be destroyed. This theory is common among all Holy Books like Quran, Gospel and Old Testament with few differences in the details here and there. Where did the Holy Books come from? Creator communicated to certain people known as "Messengers" throughout human history and delivered them the exact same message and assigned them the duty to convey this message to other humans. Whoever followed the message was saved and others were destroyed. For example Noah gave this message to the whole World which took him 950 years. After he had delivered the message or the notice to everyone God destroyed everyone except the few who believed and followed Noah. Similarly there were other messengers such as Ibrahim, Ismail, Isaac, Jacob, Joseph, David, Solomon, Moses, Jesus and Muhammad. Every messenger faced similar challenges of facing opposition and harassment from the ruling elite but ultimately they were able to succeed in their missions. Messages brought by these messengers were compiled in the form of Holy Books. Moses was given Tora or "Old Testament", Jesus was given Gospel or "New Testament" and Muhammad was given Quran which is supposed to be the "Final Testament". When and why Messengers were sent and who got to be the messenger was entirely at the discretion of God but there were clear reasons for his decisions. About half of the World population follows one of these Books and believes that the theory of God is true. Yet there are different versions and everyone believes that their version and interpretation is true. Which would be fine but somehow this difference in opinion has led to many wars and even today there are many wars being fought due to differences in religious opinions.

It appears to me that sadly all believers in God seem to be following the paths of their forefathers mindlessly and not paying attention to the details or what God actually wants from Human beings. No one is thinking and everyone is just following what they have been told by other human being. There are some important questions we need to think about. For example why God kept sending messengers one after the other but now there has not been one for the last nearly 1500 years! Why did Muhammad get to be the last messenger? Why would Jesus come back and not Muhammad? There are answers to all these questions if you try to understand the logic behind God's methods. To me it appears that no one even read Quran. It is the message from the Creator itself and it is a very important piece of scientific material and should be studied that way.

Muhammad got to be the last messenger simply because he had the Quran written down in his own life in the original form and the language of the Quran is still alive. Why would there be any reason to send another messenger when humans already have an authentic and accurate message? So when Jesus is going to come back he will not bring a new Book or message. Humans already have the right message for more than 1400 years and there is no doubt in it. And I am a witness to the accuracy and clarity of the Message. Humans do not need any new message.

Jesus had brought the same message as Muhammad did but unfortunately he never had it written down and it only started getting written down after about 30 years of Jesus being gone and hence is technically unauthentic and no one can say for sure if it does not have any inaccuracies in it. It certainly does have inaccuracies in it and that is why Creator felt the need to send another messenger that is Muhammad. He was sent to correct those inaccuracies. He confirmed Jesus to be sent by God and Gospel and Tora to be Books of God. Biggest of those inaccuracy is the fact how Jesus is given the status of a God in Gospel. Jesus had a DNA which had to be written down first before Jesus could come to life. That is what Quran clarifies in detail. Quran message to Christians is not to worship anyone but one God and do not associate anyone with him and do not take each other as lords and follow the Gospel. Quran says to Christians that if you are not upholding the law of Tora (Old Testament) and Gospel then you are standing on nothing.

Moses was also given the same message but through the history original message got altered by human manipulation and also because its language became obsolete and few of the pages from the Book were removed and some text was modified and that is why Jesus was sent to clarify those. Jesus just like Muhammad also confirmed Moses to be a true messenger and Old Testament to be the message from God. But as Christians refused to accept Muhammad as a messenger, same way Jews refused to accept both Jesus and Muhammad as messengers. How can God ever correct the man-made changes in the Books if people simply do not accept the new messengers? God says in Quran that he took a covenant from the Prophets that if another prophet comes and says that your Book is right and I am also a messenger from God to clarify few things then you must help him. And yet history keeps repeating itself and every messenger is simply denied by the established religious institutes. Now I think when Jesus will come he will likely be denied by Jews, Christians and Muslims as it has always been the case with previous messengers.

But what is important to realize is that more than the messengers we need to focus our attention on the message they had brought to us. Quran is the latest and the most authentic message (that is written format was approved by the messenger himself) and is preserved in it is original form. Quran is a message from the Creator to all the humans living on earth. It has been over 1400 years that humans have this message now and ever human by now knows that the Creator has sent this message. Now it is up to every human being to read or not to read, to understand or not to understand and to respect or ridicule this message. Muhammad did his job really well by making sure that the Book is accessible to everyone who wants it. God promise that he will deliver the message to all nations before destroying them has been fulfilled now as it was fulfilled by Noah before. Now humans have no excuse for behaving the way they are as they have the right Message and I am a witness to the accuracy of the message.

Knowingly or unknowingly more than 99% of the World population right now is disobeying the message and is in violation of the fundamental rules of God in one way or the other when you judge them against the clear criteria mentioned in Quran.

I will explain how I have come to this number and it is not that hard to understand what I am saying.

Half the population does not even believe in one God so that is an automatic failure.

Jews had gone wrong obviously long time ago. They started calling Ezra as the son of God. Old Testament says that sons of Gods were marrying daughters of humans and having children! I guess Jews of that era did not understand much about DNA. Creator does not have a DNA, it creates DNA! God sent Jesus and then Muhammad to clarify this but they still do not follow the guidance sent for them. Do they not understand that if God can send so many Prophets before Moses then why can't he send prophets after Moses? And are Jews 100% sure that Old Testament has no human manipulation?

Christians just like Jews have given a man the status of God and his son. They even know it very well that God does not forgive when you associate anyone with him or give something or someone else God's status or ascribed God powers to someone else. Muhammad was sent to clarify this to them but they have simply refused to accept him as a messenger and take Quran seriously. This is exactly the mistake Jews made. Basically they are questioning God's authority as they are not happy with the choice of the Prophet. Jesus was a man, a product of DNA. Says in Quran that Jesus example is like Adam we created him from Clay. Jesus was given a pure soul and was born a Prophet. So this tells me that he was not being tested like everyone else. But as all humans will get tested I think when he comes back he will be a regular person with a regular flawed human soul which he will purify himself. He will be tested just like every other human being and prophets who get to go to heaven. But only God will clarify that. I am just speculating here. But anyways Christians are guilty of giving a fellow human being the status of God. By the way this is not the first time Humans have done that, it has happened before too. But Christian and Jews did have a valid excuse before Quran that the Books they had were flawed. They could blame the Books. It was the Book which said Jesus was God or son of God. But once Quran was sent and this was clarified they have no excuse anymore.

Now that leaves Muslims as the ones with the right message. However when you judge Muslims against Quran, it doesn't take very long to figure out that almost all Muslims (more than 99%)are in violation of the fundamental principles of Quran. It is as if no one has read Quran at all. It is really shocking.

Here is the list of commonly accepted rules in Islam which violate the fundamental principal of the message from God and make most of the Muslim sects including Sunni, Shiite, Bora, Ismaeli and Wahabis wrong:

1. Apostasy: It means if a Muslim leaves Islam then he should be killed or punished. Most Muslim scholars believe apostasy to be part of Islam and in fact about twenty Muslim majority countries have this as law in their legislature. However all the evidence in Quran is against apostasy. Muhammad is again and again told in the Quran that his job is only to deliver the message and he is not a guardian or controller over anyone. It is clear in Quran that there is to be no compulsion in the religion. And right path has been clear from the wrong path and people must choose their own path. Let everyone choose their path. It also says that if God wanted he could have made everyone the same religion. So if Muhammad was not the guardian of anyone else's religion than who else can claim to be the guardian of anyone else's religion. People who say that Muhammad said to kill anyone who leaves Islam are basically accusing Muhammad of disobeying Allah's clear instruction. Anyone can leave Islam if they want. Why should this matter to anyone else anyways. I don't care if the whole World gives up Islam; I am not responsible for anyone else's choices. Prophets were advised to stay highly professional. Just deliver the message and then God wants to see what everyone does with that message. There is no compulsion in God's rules.

2. Blasphemy: It means if any one disrespects or ridicules Muhammad or Quran then he should be punished. Quran told Muhammad not to worry about the ridiculers.

Quran advises Muhammad that prophets were ridiculed before as well. Jesus and Moses were ridiculed too. God says clearly "that he is adequate for the ridiculers". Muhammad himself did not take any revenge and ignored the ridiculers. Now Muslims today are neither following Quran nor Muhammad's sunnah. Quran advises to ignore the ridiculers and that is what needs to be done by Muslims or any believers.

3. Killing of non-believers: **This is a major** disobedience. Quran says to offer peace when you are superior. Quran says if you are offered peace then do not fight and live in peace. Only place it says to kill non-believers is if you are at war with them. And you can only engage in war if you are attacked or oppressed such that you cannot immigrate. That is it. God does not want Muslims to go waging wars and force converting people to Islam and killing people because of their religion. What ISIS is doing cannot be far from teaching of Islam.

4. Addressing anyone but God in the prayers: Many Muslims today will address other human beings such as Muhammad, Ali, Hussein and many other local saints who have died. This is again a clear violation of the message. Allah says to only address him. It is clear in Quran that dead cannot hear and it says "you cannot make the people in the grave hear". Yet people will keep praying to fellow dead human beings. Muhammad, Ali and Hussein are all dead and they cannot hear you as per Quran. Just by simply invoking someone else one becomes guilty of association with God which is an automatic failure of the test. Do not confuse the Writer of the DNA with the product of the DNA. It's very simple. One mistake or sin which is not forgivable is that If you make the mistake of addressing someone else in your prayers.

5. Dishonesty: I have met countless Muslims who are

dishonest. If anyone thinks that they can get away with being dishonest then they have not fully understood the concept of God. There is no getting away. So dishonest person is not a true believer in my opinion. A true believer knows that this life is just a test and everything is being recorded and there will be hell to pay.

6. Reading Quran without understanding the meaning of it: **Anyone who reads Quran in Arabic without knowing Arabic is not learning anything from Quran at all. It is so dumb but sadly true that most of people in Pakistan including thousands of religious Madrassa teach Quran only in Arabic where people don't understand Arabic. People understand the message! You cannot understand anything if you don't know the language. Quran has been made available to you in your language so read it in the language you understand.**

7. Making distinction among Prophets: **Muslims believe Muhammad to be superior then others prophets. Where Quran clearly says not to make any distinction between any of the Prophets. All Prophets such as Muhammad, Jesus, Moses and Ibrahim etc are equally respectable for every human being. Only God has the authority to make such distinction. I personally think it will be Jesus who will get to be the King in heaven given he will have the knowledge of a modern man which would be clearly superior to any other prophet. After all it was human superior knowledge that gave human a higher status than angels. However I could be wrong and I will happily accept whoever God picks as the best Human being ever made.**

8. Getting divided into sects: **Quran says not to get divided into sects and yet Muslims have divided themselves into so many sects. Quran also says that people who get divided into sects will taste the violence of each other and this is exactly what** is happening right now to most of the Muslim World. About 8 different Muslim countries are at war with each other and are tasting violence of each other. Quran says to follow the path of Ibrahim and do not get divided.

Quran advises Muslims, Christians and Jews and basically all humans to follow the path of Ibrahim. This is the same path shown to us by every Prophet. Moses, Jesus and Muhammad all walked in the footsteps of Ibrahim. Sadly this path is very lonely right now. I might have met perhaps one or two people who were looking for this path just like me. May Allah help them find it. Rest of the World including believers and non-believers alike has gone somewhere else in completely wrong directions and don't even realize or care about it. I am not even sure if any of them will ever see how wrong they have gone. Emotions and egos have blinded almost every human being on this planet right now. Usually these are the times when a prophet is sent. And as this time it will be the last Prophet Jesus and with him this World will come to an end. If I am right, planet Earth does not have much time left. I often think what Jesus would do when he shows up. He has some work cut out for him, doesn't he? Christians expect him to come as a god, flying with a sword in his mouth going around killing people! Quran says he will be a human being and will have "knowledge of the hour". So I think this time Jesus would be just like anyother prophet. A regular man who does not know he is supposed to be Jesus till God assigns him the duty to save the World and be the witness over it. Again he will not be bringing any new Book or detail message as humans already have Quran. How he will get his job done without getting killed I dont know but I am curious. He will have to be smart about it. Any ways I think Jesus will show up soon and World is about to end as promised by God. I don't think it will take long for Jesus to deliver a last warning to humans thanks to internet and TV. So don't think you will have a lot of time. Try to become the logical and good hearted human being God wanted to create. Picture yourself in a role where you are responsible for all other creatures on Earth; God successive authority. Elevate your status such that God can proudly ask the angels to prostrate to you. Only then you are worthy of going to heaven.

May Allah correct me if I have gone wrong. I am not walking behind anyone; I have found my own path by reading the Message (Quran) sent to me By Allah through Muhammad. And this is my message to every human being on this planet. "You don't know how much time you have got but I have a feeling that time has run out. May Allah guide everyone who looks to the path of Ibrahim, Moses, Jesus and Muhammad. It is the one and only straight path which goes back to Allah/God/Creator. He has all these names. Quran has the right guidance for all humans if anyone wants to be guided".

In summary once again the message of Quran is that there is only One God, only pray to him and he knows everything you are doing and why you are doing it. You cannot fool him and if you try then know that you are only fooling yourself.

Be honest, always seek the truth, be kind, helpful, knowledgeable and charitable and stand up for justice even if it's against yourself. If you have still not read and tried to understand Quran then it is your own fault as it has been made available to everyone in all languages now. If you don't speak Arabic then read or listen to Quran in a language you can understand and pay attention to details and do not and I repeat do not doubt the Book as there is no doubt in Quran. If you read it with doubt in your heart then you will be misguided and go astray and that is how the Book differentiates the true believers from the hypocrite and non-believers. And if you accept Quran to be true and accept what it wants you to accept then do not think that you have done me or God any favor. You have simply done a favor to yourself and have saved yourself form living in hell forever. No one else benefits or gets harmed by your accepting or rejecting the message respectively except you yourself and that is clear in Quran. No one can force you to accept or reject the message at any time. It must be your own choice to accept or reject and Allah wants to see what you choose. Most of the people on this earth are wrong and if you follow anyone you will go wrong too. So read Quran by yourself and find your own path. If Creator sees any good in you, he will guide you and will not let you go astray. Make your sincere effort and understand the concept in the Message. Peace!

Governing Structure - Equality vs. Inequality

The fate of a nation lies in the hand of its leader. Who gets to be the leader is a very important decision and deserves some attention. Most of the civil World believes in selecting the leader based on majority opinion. This form of governing structure is known as Democracy.

Democracy is defined as "government of the people, by the people for the people". It is a neat concept and it solved many of the problems which kingdoms had faced where son of a king was a king and ultimately kings became above the law. Democratic leaders are not above the law, they have to perform and be transparent or fear losing the next elections and persecution. Everyone in the society is treated equally and has been given an equal say or worth towards choosing the leader. Today "democracy" is considered the best form of government by most civil countries in the World and more and more countries are trying to adopt this way of government.

However democracy has certain flaws not only in its concepts but also in its feasibility which essentially makes a democratic government helpless. One has to look at it very closely to understand how democracy works in order to understand how it actually does not work. Though these flaws are not obvious but they are not small and in fact could make a democratic government worse than a kingdom in certain situations.

Democracy assumes few assumptions which are simply not true. One assumption is that every person's opinion has the same value towards choosing the next leader. Which one acquires when one reaches the age of 18 years. That is the only criteria; no attention is given to any other variables which enhance someone's wisdom and value of one's opinion such as age, experience, education or intelligence etc. So an 18 year old high school dropout and a 50 year old political science professor get assigned the same worth to their opinions. Basically you assume that professor never went to school and he has nothing additional to contribute over an 18 year old kid. You assume that professor mind stopped growing at the age of 18 and he has not learnt anything in life since then which is clearly not the case. But why no one objects to it? How could smart people like Einstein simply accept the fact that his opinion has the same worth as a high school drop out? The reason is "a vote" has been given the status of a "basic" human right. And as all humans are equally human then everyone gets same rights. So real question is, is the vote a birth right or a hard earned worth of opinion? Say if I did not vote will I not get access to health care and food! That should not be the case and luckily it is not. Basic human rights are what you should get for being just a human and those are equal access to food, education, health and justice. What you are able to contribute or how valuable your opinion should be are not basic rights. Do educated and uneducated have the equal ability to judge? Einstein and a high school dropout both should get equal access to justice but their opinions on who should be the next leader by no means should be treated equally. Value of one's opinion is not the same for everyone. This is distinction which is a hard earned status. Are all fish equal in their worth? No, because their value or worth lies in their meat. Bigger the fish more valuable it is. Human worth lies in their brains and opinions reflect that worth. Not all humans have the same worth and we accept this in our everyday life as everyone is paid differently based on their experience and education etc. We are equally humans as far as basic rights are concerned but our opinions have different worth. And enforcing that everyone's opinion will be treated equally is simply illogical as there are clearly superior and inferior opinions. You surely hope that an educated person opinion about most things in this World is going to be better than an uneducated individual. And if it is not the case then there is something seriously wrong with your education system. And obviously age and experience have huge impacts on one's outlook. I am not getting educated anymore but every day I live I learn something new. My choices and opinions would be clearly different now than when I was 18 because the variables which I based my choices and opinions upon have completely changed. Thank God they have changed. I am not an 18 year old kid anymore. If I did not then there would be a serious problem. But yet my opinions are still treated the same by democracy as if I am still 18 year old. Where is the logic in this? There is no logic. Truth is "a vote" is not a basic human right. It is the worth of your opinion towards selecting a leader of a nation and logic will tell you that you simply cannot treat all opinions the same as they simply are not the same.

Basically democracies count the bodies and do not look at the mind where in fact human worth lies. Humans are only born equal but then some get education and grow but some don't and this differentiation is very important and not accounting for this when assigning worth to a human is a fundamental flaw in democracy. It's like having a fishing derby where fish are not weighed but simply counted with an assumption that "all fish in the ocean are equal in worth". So one can catch a world record blue whale but will be beaten by someone who catches two minnows. The truth is not all humans' opinions are equal in their worth and developing a system based on a wrong assumption is only going to take you in the wrong direction. Superiority in opinion is a hard earned distinction earned by a certain minority by dedicating their lives to knowledge and experience and hence gaining wisdom. Inequality of opinion should be embraced and cherished. You are simply letting it drowned by the not so distinguished majority.

Say you are educated and your time is paid different than a high school dropout so how come you are stripped of this entitlement when it comes to weighing in on how your society should be governed. You made certain choices in life to get where you are and the society should give you more credit as you are in better position to chose between right and wrong. Say if humans are like rocks, some are gravel stones, some are rubies and granite and few are rare diamonds. Though these are all rocks, they are not equal in worth. It's only wise to differentiate among the rocks no matter how difficult of a task that is. You will have to assign this job to people who know about rocks and minerals and can spot a diamond when they see one. Simply saying all rocks are equal and let's just count them is not wise at all.

In democracy your individual net worth is not "one vote" as you are told. Your true or net worth depends on the overall number of total population in a voting unit. Say if there are 100 people deciding who gets to be the leader then your worth is 1/100 and if there are 100,000 people than your worth is 1/100,000. So the smartest and most educated person in California has a lot less worth (1/15,000,000 vote) to his opinion then a college drop out in Wyoming (1/270,000 vote). And on top of that when everyone is basically given the exact same worth then essentially no one gets any worth. The worth or value of anything is derived by inequality itself. This rule applies to power of anything including electric current. If both wires had 110 volts then there would be no useful power in both of them. Worth or power is defined by inequality itself. So democracy gives an illusion of worth but in reality it takes worth away from everyone. In America there are about 146 million registered voters. This gives every Americans' vote of an overall worth of 1/146,000,000 vote = 0.000000006 vote. Many people subconsciously realize that and hence do not bother voting at all. No matter how smart or knowing you are you have the same insignificant value.

Why is everyone assigned an equal worth is another question? Is it an intentional scam or just a plain overlooking of the simple truth? I don't know the answer to that but I would like to give the benefit of doubt and believe people never thought about it objectively before.

Now let's look at another aspect of democracy. Should everyone be involved at all in the selection process of the leaders? Typically hiring a candidate for a job is a task assigned to only who are in position to understand the job requirements and all applicants' strengths and weaknesses. And that is the only logical way. Why would you ask someone to hire for a job when that person does not understand the job requirements or the applicants resume? Then why do people insist on being involved in this selection process when they don't understand even the job requirements. Leaders are supposed to be the wisest and the most sincere among the masses and it takes wisdom and sincerity to recognize and acknowledge those qualities. For example if you have bunch of diamonds and you want to pick the best diamond then would you not value a jeweler's opinion over any 18 year old kid's opinion? Selecting a leader in a country is just like finding the best diamond. You ask an expert jeweler who is able to judge diamonds. Don't just ask random people to vote which is the best diamond. They will never pick the right stone as most of them can't even tell a diamond from a junk piece of glass. Democracy will bring out the most popular image but not necessarily the wisest or the most sincere. Hence G.W. Bush in America and Asif Ali Zardari in Pakistan got elected as "leaders" of their nations. I am sure there were smarter people then Bush to run the country. You gave power to someone like Bush and he destroyed the whole World by starting all the wars which are still going on and now no one knows how to stop them. Zardari came into power in Pakistan and the country got destroyed. If people like Zardari can become leaders of any nation then whatever process they came through is highly flawed by definition and should be abandoned. You don't need any further evidence against democracy than Asif Ali Zardari becoming the president of Pakistan. Case closed.

Now let's look at the things from the elected "leaders" point of view. First of all democratic leaders shouldn't be called leaders at all. They are more like hired drivers or employees or "followers". A follower simply cannot lead as a principal. Let's call them "hired drivers" of the bus you are all in it. If the majority tells them to go right then they have no choice but to go right even though the right direction might be left. Leaders are supposed to be the wisest and lead the way and not follow the orders from the not so wise majority of the bus. But the "hired drivers" have to face a major conflict of interest in all their decisions; appease the public and their donors or do the right thing and often it's not the same choice. Ultimately a hired driver has no choice but to simply obey the public demand and their donors' interests in favor of what is the right thing to do. Most of such drivers are not even capable of recognizing what is the right direction in first place. They simply are there to do what majority says to do. Right and wrong does not matter, what matters is what public wants. Keep the public happy and stay in power.

Otherwise one would risk losing a certain vote bank. Fact is there is no scientific correlation between the truth and how many people say what truth is. But in democracy world if a certain majority says it's the truth then it becomes the truth. In fact for a truth to be considered a truth a certain majority has to say it. That's why you protest, bigger your protests, louder your voices, more right you are. It's a pack mentality. Protesting and for that reason whole concept of democracy is a form of bullying based on the strength of your pack. If I have 100 people behind me and you have 50 then I am automatically right. Numbers matter in this system of government just like they do in pack animals. In a pack, all animals are more or less look the same, weigh the same, sound the same and have the same small brain and hence the same worth and are truly equal in value. So numbers to a pack matter. But humans are not designed to be pack animals and we are not all equal. Humans made tribes and a tribe leader was the wisest and the strongest man who could ensure survival of the tribe. 18 year old kids did not decide who got to be the leader. A sincere tribe leader is like a father of the tribe. He typically would not get directions about where to go and what to do from the kids. He would lead them and everyone simply trusted him and followed him. If he was not qualified to do so then the tribe was doomed. And hence nations like Pakistan where people like Asif Ali Zardari and Nawaz Sharif become the leaders get destroyed completely by democracy.

Question is if people don't have much say individually and the drivers don't have much control then who is really in control? Answer is "majority" and you all know that majority is not as qualified as one would hope. It is an assumption that majority knows and can decide which is the right direction. Or to be specific 18 year olds get to decide where the bus should go and not the professors as they are outnumbered. So democracy defies logic. Logic is what makes you human. Be human, use logic and embrace inequality of opinions.

Another assumption in democracy is that a leader exists in a society. In fact many times there is no able leader in a certain society. Like sometimes a CEO is hired from outside, I think a leader could be imported too as long as he is sincere towards the nation. And even if say there is a true wise leader living in a society, democracy assumes he will also step forward and join the beauty contest or the popularity circus whatever you want to call it. And another assumption is he/she will also win in the popularity circus. These are all false assumptions. In reality democracy will bring out the most popular image and not the wisest character. To be popular requires completely different skill sets then to be a leader. Good leaders are not necessarily the most popular as they are typically not people pleasers or good at making deals with the investors.

Another major problem with democracy application is evident in countries which have not secured fair justice for all. Over there humans are considered just a vote and not humans. And the "leaders" job is to secure the vote no matter what. So this vote becomes a collar in dog's neck via which they are yanked around by people in power. If you don't vote for the right person then you better be ready to deal with consequences. I have witnessed this first hand in Pakistan. Poor people never get to exert their free will for example and even if they did what choices do they have to choose from. And even if there was a clear choice they are unable to recognize it as they don't understand the job requirement or candidates resumes etc. None of the villagers for that matter can vote freely, they are bullied, harassed and manipulated into voting where the dominant figure of the area orders them too. And if they are caught voting with their free will then they are harassed often by police itself who is only there to guard the interests of people in power. Say if I wanted to run for a political position, no matter how honest, qualified and sincere I am at the beginning of my campaign, I would be a different kind of person by the end. If I am serious about winning then I will have to make deals with the thugs and mafia who have accumulated a certain vote bank without asking how they control so many votes. I will have to turn a blind eye to all the oppression and injustice or I simply won't get the votes needed to win. And after losing I will be locked up or harassed by police for daring to try to oppose an already established and respectable political personality. A good example is Imran Khan in Pakistan. The World cannot find a more sincere and qualified leader than Imran Khan and yet for last twenty years he has not been given a chance. Every crook in the country got into power but him despite everyone including the military and the international community acknowledging that he is the best person to lead the country. This is the sad reality of democracy.

Democracy has lead to destruction of many nations. One perfect example is Pakistan who has been obsessively trying to achieve an ideal democracy and in this pursuit has brought the country on a verge of complete collapse. Democracy has become an industry controlled by political mafia. It assumes humans are nothing but equally worthless gravel rocks or fish; whoever can collect more rocks or can catch "more" fish gets to lead. Especially dangerous is where justice system is not fair and available to everyone. The combination of injustice and democracy is highly oppressive. And democracy can never fix injustice as injustice always brings corrupt democracy and this becomes a vicious cycle from which countries can never get out of unless they let go of democracy. Fish in a fishing derby are treated more fairly for their worth than humans in a democracy. Human worth is in their brains and not in bodies but yet democracy completely ignores the brains and insists on counting the bodies. As long as you can breathe your worth is same as anyone else. No its not!

Here is an example, imagine what would happen to any country military if the military decided to pick its chief based on a democratic process! It would totally destroy the order and turn military into a circus as well. Same thing happens to a country before any elections. If you don't agree with me just follow what is going on in America right now. In Guatemala a comedian got elected as a president. That is not a joke!

Next question is what the alternative is if democracy is not a viable option. You need to backtrack and learn from the biology. You are not pack animals and not created equal and only the ones wise enough to understand the job requirement and the candidates' resumes should be involved in selecting a leader based on some merit and not just popularity. Humans lived in tribes and successful tribes would select a leader based on merit and only the wise elders got to decide who the leader is going to be. A leader had to be strong, smart and most importantly sincere to its people as the whole tribe survival depended on him. You need a tribe leader, a king or CEO. A CEO who is sincere and the wisest and is chosen by a board of sincere trustees should be allowed to lead but he should be subjected to same law and justice system as everyone else. He may not be the most popular but he doesn't have to be. He has to treat all humans like his family. A father is a king of his family. There is no democracy in a home. The difference between an oppressive dictator and a beloved king is the same as a drunken abusive father and loving dad. One drunken father doesn't mean all fathers are abusive. World should look for a father who is willing to be firm when needed and give his life for his family if needed. America got lucky with George Washington, as he truly was a father. He was more like a king but he made the mistake of giving the illusion that he came via democratic process. Such men are hard to find and if the world finds one it should hang on to the same king as long as possible and as long as the king remains capable and sincere to his people.

Qualities of a king should be:

1. Education – very important factor, it should be broad based education with emphasis to sciences, religions, environment etc

2. Wisdom - knowledge is not the same as wisdom. A wise king knows when and where to apply his knowledge

3. Humility - it comes from knowledge and wisdom. More one learns more one knows that he doesn't know anything

4. Health. King is a role model. He should be healthy and fit

5. Sportsmanship. A very important characteristic for any leader

6. Selflessness. A king should be able to put his people before his own interest

7. Fearlessness. Intimidation will be part of his life. So being intelligent and physically strong makes one less vulnerable. And if one is sincere in one's intentions than he has nothing to fear from anyone else.

8. Afraid of a higher power. This is the most important quality. Because a king who is truly afraid of God knows in his heart that no matter how much power he has, he has no power and he is never alone and he can't get away with anything at all. A truly God fearing king cannot be corrupted by definition. If he gets corrupted then he did not believe in God to begin with and was simply a hypocrite. Power corrupts and absolute power corrupts absolutely and all humans are vulnerable to this unless they believe that all

power belongs to God. Such king will be accountable for this responsibility of a king assigned to him.

If you continue with democracy, sooner or later it will divide you among smaller and smaller groups who will be trying to get bigger and bigger. This phenomenon is already obvious even in the most civilized countries who claim to be champions of democracy. Sooner or later the World will have to get rid of democracy and the notion that everyone has equal worth. Sooner you do this the better chances you have. Doomsday clock has been moved already to 11:57. You are all equally human beings but you don't have the same wisdom which comes from knowledge, experience, age and intelligence. You should start respecting this simple fact of life. Otherwise you are heading towards more chaos and destruction. If you want to protect your gene pool you need to change the ways you govern yourselves. It's not a small change I understand but it's based on logic. And if the humans still have some ability to reason left in them then I am hopeful that many will understand my point of view. But again how many people need to understand my point of view in order to bring this change? Say some other smart person proves me right and agrees that current democracy is taking you towards destruction, will the world change then? How do you change a system based on majority opinion against the majority opinion? So really truth does not matter at the end. This was Galileo's dilemma. He was outnumbered and he could not change majority's opinion and majority won by bullying him into changing his opinion. No matter how many people say against it, a truth simply remains the truth whether majority likes it or not.

Based on my observations I have come to this conclusion that none of the World major problems will get solved as long as the World continues to choose leaders based on general public opinion. Sadly world is so fixated about it. Every day I see another nation opting for democracy. Watch what happens to the Middle Eastern and African nations who have recently celebrated democracies. You can learn a lot from their destruction. This is why I have denounced democratic process and have vowed never to vote again. I am not just a warm body that I should be counted. As long as there is justice being served in any country that country will be okay. There is justice in Canada and that is why it is peaceful here. It is justice which guarantees freedom and maintains peace not democracy and they are not the same thing. Every country needs to focus on providing equal justice for its population. Who gets to select the leader is less important in the end as long as you get a God fearing, sincere and wise leader who does not have to worry about securing votes and investors for his party's future. A king doesn't have to be Mr. Know-it-all. A sincere king will find other sincere, capable and knowing people and put together a team to pull his people out of the chaos and bring peace, justice and prosperity to his nation.

My message to the military chiefs of all the countries in the World is to take control of their countries immediately. Once in control look for and sincerely select a leader for their nations based on merit and criteria outlined above. If the military chiefs cannot get sincere towards their nations than those nations are doomed. But know that whoever created you is watching you closely. And I will be a witness against you on the day when all of you will be asked about the responsibilities assigned to you. Know that hell is not just a story. Once a capable and sincere king has taken oath, his first duty should be to bring justice for all, freedom of speech and freedom to choose and practice a religion and abandon a religion. All humans are created equally. You are all equally Homo sapiens including the king. If a king fails to bring justice for all within three months then he should be replaced.

My message to Gernal Raheel Shareef in Pakistan is to get rid of of Nawaz Sharif today and make Imran Khan the king of Pakistan. That is the only way you can save Pakistan from the crisis she is in. You can laugh and ridicule this idea as much as you like but sooner or later you will have to agree with me. If the end goal is saving Pakistan then this is the only option. Pakistan is lucky that she has a capable and sincere leader readily available. Not all nations are going to be this lucky.

As far as Canada is concerned I can see Justin Trudeau as making a good king for Canada. He reminds me of Imran Khan. They are both courageous, upright, educated and sportsmen. I hope he is as sincere for Canada as Imran Khan is for Pakistan. Sheikh Muhammad of Abu Dhabi has to go. He is guilty of buying, starving and abusing children to be used as camel jockeys. King of Saudi Arabia has to go as he is simply not qualified enough and has failed to bring justice for all in Saudia Arabia and has failed to understand Islam. In fact all kings and queens who were simply born kings or queens without meeting any criteria have to go. No one owns a country so they can simply inherit it. People own their countries and everyone deserves a qualified and sincere king who could ensure justice, peace, food, health and education for all equally. If the military chiefs are happy with the current leaders as kings then its up to their discretion to keep them or not. But get rid of democracy circus and get serious about saving your nations and the World. But then what do I know about politics, I am just a doctor.

CPSIA information can be obtained
at www.ICGtesting.com
Printed in the USA
LVOW01s1919201115
463393LV00021B/79/P